LAW OF HEALTH AND SAFETY AT WORK: THE NEW APPROACH

AUSTRALIA
The Law Book Company Ltd.
Sydney : Melbourne : Brisbane

CANADA AND U.S.A.
Oceana Publications Inc.
New York

INDIA
N.M. Tripathi Private Ltd.
Bombay
and
Eastern Law House Private Ltd.
Calcutta and Delhi
M.P.P. House
Bangalore

ISRAEL
Steimatzky's Agency Ltd.
Jerusalem : Tel Aviv : Haifa

MALAYSIA : SINGAPORE : BRUNEI
Malayan Law Journal (Pte.) Ltd.
Singapore

NEW ZEALAND
Sweet & Maxwell (N.Z.) Ltd.
Auckland

PAKISTAN
Pakistan Law House
Karachi

LAW OF HEALTH AND SAFETY AT WORK:

THE NEW APPROACH

by

CHARLES D. DRAKE, M.A., LL.B.

of the Middle Temple, Barrister
Professor of English Law, University of Leeds

and

FRANK B. WRIGHT, LL.B., F.R.S.H.

Senior Lecturer in Law, School of Law, Leeds Polytechnic

With a Foreword by

JOHN H. LOCKE, M.A.,

Director, Health and Safety Executive

LONDON
SWEET & MAXWELL
1983

Published in 1983 by
Sweet & Maxwell Ltd. of
11 New Fetter Lane, London
Computerset by Promenade Graphics Ltd., Cheltenham
Printed in Scotland

British Library Cataloguing in Publication Data

Drake, Charles
Law of Health and Safety at Work.
1. Great Britain. Health and Safety at Work etc. Act 1974
2. Industrial safety—Law and legislation—Great Britain
3. Industrial hygiene—Law and legislation—Great Britain
I. Title II. Wright, Frank
344.1041465 KD3168.A3/

ISBN 0-421-28620-2
ISBN 0-421-28630-X Pbk

FOREWORD

Eight years have now passed since the Health and Safety at Work Act 1974 came into force. Parliament intended that the new Act should be the basis for major changes in the form and substance of legislation to improve standards of health and safety. We can now see the way in which Parliament's intentions are being fulfilled.

The general duties set out in the 1974 Act require everybody whose activities could create hazards to workpeople or the general public to do all that they sensibly can to see that people do not get hurt as a result. It is essential for anyone running a business or any other organisation to understand clearly what this involves. No longer is it possible to think of health and safety obligations as simply complying with a mass of detailed requirements set out in old Acts or Regulations.

The 1974 Act also set up new institutions and it is important that those responsible for health and safety should understand how these are working in practice. Regulations and Codes of Practice are now being prepared and approved through the Health and Safety Commission which is representative of the main interests concerned—employers, trade unions, and local authorities with their concern for the interests of the general public. A complex system for consulting those concerned has been developed and it is important that everyone should understand how that system works.

The Act also set up the Health and Safety Executive as the main enforcement authority using new enforcement techniques. The Executive wants those with whom it deals to understand the way in which Inspectors and others operate the new legislation. We are only at an early stage in working out the changes which the 1974 Act began. A new pattern of Regulations and Codes of Practice is steadily emerging to deal with major areas of concern. Much of the existing detailed legislation will, I am sure, be gradually replaced. All concerned with health and safety need to keep abreast of the changes which affect them. But the main framework of the 1974 Act seems to be standing up well. It will be most effective if all with responsibilities for health and safety are clear both about the general philosophy and the way it operates in practice.

I hope this new book on health and safety at work law will help towards a fuller understanding of these important changes.

October 1982 John Locke

PREFACE

In this book we attempt to assess within a manageable compass the new system for health and safety at work introduced by the Health and Safety at Work etc. Act 1974. That Act—or "super-Act" as we have termed it—did not come a moment too soon to meet the challenges presented by new technology, new substances and processes and new dimensions of scale in the industrial process, all of which have brought in their train new or increased risks. These structural changes have been accompanied by social change. People are better informed and make greater demands on society than their forbears. Unsafe or unhealthy conditions at work which were stoically accepted by previous generations are increasingly questioned by workers whose commitment to the "work ethic" lacks the theological fervour of the Victorians. At an early stage we discussed the possibility of trying to accommodate treatment of the main protective "codes" relating to factories, mines and quarries, agriculture, offices, shops and railway premises but decided against this in favour of concentrating on the post-1974 system. Quite apart from the well-nigh impossible task of accommodating any satisfactory treatment of the old laws in a work of reasonable proportions, we wished to look at the new system as it is, and as it is likely to develop, rather than to look back to the old protective laws, even though it now seems unlikely that these will be replaced as speedily as was at first envisaged.

This approach has enabled us to consider the health and safety system against the background of a common law whose aim is compensatory and to examine the role of the law in relation to problems of values and reform. In the treatment of the law itself we have sought to eschew any holistic adumbration of rules but to place those rules within the social, historical and political environment within which they operate. We have described events and institutions when to do so will lend interest to comprehension of the law. In doing so we lay no claim to polymath abilities but we are aware of the risk that the legal specialist may obtain a distorted picture through his legal telescope. Health and safety at work is the meeting place for several disparate disciplines. Scientific research, economics and engineering may have more to offer to the solution of a health or safety problem than a reading of the relevant legal rule. The determination of threshold limits, the application of cost-benefit analysis, or the use of probability measurements, can serve to illuminate the application of purely legal norms. The lawyer, however, must not be too self-abnegating since the system has to operate within what ultimately is a legal framework. That framework has inevitably been affected by the increasing pressure for harmonisation of laws from abroad. The European Communities and, to a lesser extent, the International Labour Organisation, require our national law to be examined for conformity with these supra-national health and safety norms.

We have tried to take account of these international sources, including the United States of America to whose Occupational Safety and Health Act of 1970 we occasionally make reference.

Responsibility for the text is joint in the fullest sense of that word. We are indebted to those who set aside valuable time from their own pressing duties to set us right on certain parts of the book. Professor Brian Hogan kindly cleared up a few difficulties which we experienced with that avowedly practical but essentially abstract subject—criminal law. Our thanks are also due to Frans van Kraay of Leeds Polytechnic and to Dr. W. J. Hunter of the EEC Commission for their invaluable assistance with the section on the European Communities. To these and others whose help we acknowledge we extend the customary dispensation from responsibility for such errors as may remain despite their assistance. No doubt, our readers will set us right where they think we have fallen into error; we welcome this since we can all profit from correction. We acknowledge with thanks the secretarial help which we have received, particularly from Mrs. Doris Jones (in the year of her retirement from the Law Department of the University of Leeds) and also from Mrs. Christine Taylor. Finally, we record with gratitude the support and assistance readily provided by the publishers, Sweet and Maxwell, from the initial mooting of the idea of a book on this subject to its completion.

CHARLES D. DRAKE
FRANK B. WRIGHT

Leeds, April 30, 1982.

CONTENTS

TABLE OF CASES

Adamson *v.* Jarvis (1827) 4 Bing. 66; (1827) 12 Moo.C.P. 241; (1827) 5 L.J.O.S.C.P. 68; (1827) 130 E.R. 693 158
Adsett B.K. *v.* K. Steelfounders and Engineers Ltd. [1953] 1 W.L.R. 773; (1953) 97 S.J. 419; [1953] 2 All E.R. 320; (1953) 51 L.G.R. 418, C.A.; affirming [1953] 1 W.L.R. 137 63
Aitchison *v.* Doris (Howard) Ltd. [1979] S.L.T. 22.......... 82, 86, 88
Anns *v.* Merton L.B.C. [1978] A.C. 728; [1977] 2 W.L.R. 1024; (1977) 121 S.J. 377; [1977] 2 All E.R. 492; (1977) 75 L.G.R. 555; (1977) 243 E.G. 523, H.L.; *sub nom.* Anns *v.* Walcroft Property Co. (1976) 241 E.G. 311, C.A. 159
Arkin *v.* London and North Eastern Railway Co. [1929] All E.R.Rep. 65 180
Armour *v.* Skeen [1977] I.R.L.R. 310; 1977 J.C. 15; 1977 S.L.T. 71 .... 12, 170, 215
Associated Dairies Ltd. *v.* Hartley [1979] I.R.L.R. 171 .... 66, 107
Association of Scientific, Technical and Managerial Staff (T.C.O. Section) *v.* Post Office [1980] I.R.L.R. 475 193
Automatic Woodturning Co. Ltd. *v.* Stringer [1957] A.C. 544; [1957] 2 W.L.R. 203; (1957) 101 S.J. 106; [1957] 1 All E.R. 90; (1957) 55 L.G.R. 77, H.L.; reversing in part and affirming in part [1956] 1 W.L.R. 138 111

Baldwin & Partners *v.* Brazendale, 16.3.77 (HS 42194/76).......... 138, 142
Barkway *v.* South Wales Transport Co. [1950] A.C. 185; (1950) 66 T.L.R. (Pt. 1) 597; (1950) 114 J.P. 172; [1950] W.N. 95; (1950) 94 S.J. 128; [1950] 1 All E.R. 392 71
Bartlett *v.* Newble (1979) (HS 18046/79) 68, 144, 153
Belhaven Brewery Co. Ltd. *v.* McLean [1975] I.R.L.R. 370.......... 66, 67, 148
Benn *v.* Kamm and Co. Ltd. [1952] 2 Q.B. 127; [1952] 1 T.L.R. 873; [1952] 1 All E.R. 833, C.A.; reversing [1952] 1 T.L.R. 57 111
Bewlay Properties Ltd. *v.* Jackson, 7.12.77 (HS 7/77—Scotland) 148
Biddle *v.* Truvox Engineering Ltd. [1952] 1 K.B. 101; [1951] 2 T.L.R. 968; (1951) 95 S.J. 729; [1951] 2 All E.R. 835 93
Birmingham and Midland Motor Omnibus Co. Ltd. *v.* London and North Western Railway Co. [1913] 3 K.B. 850; (1913) 83 L.J.K.B. 474; (1913) 109 L.T. 64; (1913) 57 S.J. 752 180
Blocking Services Ltd. *v.* North, 25.10.78 (C.O.I.T. HS 2/30) 148
Booker Wellman Ltd. *v.* Micklethwaite, 28.11.78 (C.O.I.T. HS 2/46) 150
Bottomley *v.* Fellowes, 9.5.80 (HS 2/201) 147
Bourne Chemical Industries Ltd. *v.* Affleck, 9.11.79 (HS 2/147) 153
Bouzourou *v.* Ottoman Bank [1930] A.C. 271; (1930) L.J.P.C. 166; (1930) 142 L.T. 535 23
Box *v.* Ware, 7.5.81 (HS 7446/81) 152
Boyton *v.* Willment Bros. Ltd. [1971] 1 W.L.R. 1625; (1971) 115 S.J. 673; [1971] 3 All E.R. 624, C.A.; reversing (1970) 114 S.J. 972 8
Bradford *v.* Robinson Rentals Ltd. [1967] 1 W.L.R. 37; (1967) 111 S.J. 33; [1967] 1 All E.R. 267; (1967) 1 K.I.R. 486 74
Bramham *v.* Lyons (J.) & Co. Ltd. [1962] 1 W.L.R. 1048; (1962) 106 S.J. 588; [1962] 3 All E.R. 281; (1962) 60 L.G.R. 453 65
Bressingham Steam Preservation Co. Ltd. *v.* Sincock, 5.4.77 (HS 27434/76) 153
Brew Brothers Ltd. *v.* Mallon (HS 17106/76/A) (unreported) 69
Brewer *v.* Dunstan, 18.10.78 (C.O.I.T. HS 2/20) 151
British Aircraft Corporation *v.* Austin [1978] I.R.L.R. 332.......... 25, 70
British Airways Board *v.* Henderson [1979] I.C.R. 77.......... 140, 146
Broad Sawmills Ltd. *v.* Radcliffe, 27.7.79 (C.O.I.T. HS 2/114).......... 139, 147
Bromley L.B.C. *v.* Greater London Council [1982] 1 All E.R. 129 12
Brown *v.* National Coal Board [1962] A.C. 574; [1962] 2 W.L.R. 269; (1962) 106 S.J. 74; [1962] 1 All E.R. 81, H.L.; affirming [1961] 1 Q.B. 303 67

—— *v.* Stevenson, 30.3.78 (HS 2/78) 149
Browning *v.* Crumlin Valley Collieries Ltd. [1926] 1 K.B. 522; (1926) 95 L.J.K.B. 711; (1926) 134 L.T. 603; (1926) 90 J.P. 201; (1926) 42 T.L.R. 323; [1926] All E.R.Rep. 132; (1926) 24 L.G.R. 302 .... 22, 31
Bullard *v.* Croydon Hospital Group Management Committee [1953] 1 Q.B. 511; [1953] 2 W.L.R. 470; (1953) 97 S.J. 155; [1953] 1 All E.R. 596 159
Bux *v.* Slough Metals Ltd. [1973] 1 W.L.R. 1358; (1973) 117 S.J. 615; [1974] 1 All E.R. 262; (1974) 15 K.I.R. 126; [1974] 1 Lloyd's Rep. 155, C.A.; reversing in part (1973) 14 K.I.R. 179 71

Cadman *v.* Johnson, 29.1.81 (C.O.I.T. HS 3/18) 149
Callow (F.E.) (Engineers) Ltd. *v.* Johnson [1971] A.C. 740; [1970] 3 W.L.R. 982; (1970) 114 S.J. 846; [1970] 3 All E.R. 639; (1970) 10 K.I.R. 35, H.L.; affirming [1970] 2 Q.B. 1 7
Campbell *v.* Wallsend Slipway and Engineering Co. Ltd. [1978] I.C.R. 1015; (1977) 121 S.J. 334; [1977] Crim.L.R. 351 .... 129, 169
Campion *v.* Hughes [1975] I.R.L.R. 291 .... 154, 156
Caravan Parts (Supply) Ltd. *v.* Peacey (HS 11776/80) (unreported) 69
Cartwright *v.* G.K.N. Sankey (1973) 14 K.I.R. 349, C.A.; reversing (1973) 116 S.J. 433; (1972) 12 K.I.R. 453 76
Central Asbestos Co. *v.* Dodd [1973] A.C. 518; [1972] 3 W.L.R. 333; (1972) 116 S.J. 584; [1972] 2 All E.R. 1135; [1972] 2 Lloyd's Rep. 413; (1972) K.I.R. 75, H.L.; affirming [1972] 1 Q.B. 244; [1971] 3 All E.R. 204 15
Central Tyre Co. (South Side) Ltd. *v.* Ralph, 15.9.78 (HS 17863/78) 58, 149
Charles *v.* Smith (S.) and Sons (Engineering) Ltd. [1954] 1 W.L.R. 451; (1954) 98 S.J. 146; [1954] 1 All E.R. 499; (1954) L.G.R. 187 105
Cheston Woodware Ltd. *v.* Coppell, 26.7.79 (C.O.I.T. HS 2/110) 147
Chethams *v.* Westminster City Council, 13.12.77 (C.O.I.T. HS 1/113) 149
Chipchase *v.* British Titan Products Co. Ltd. [1956] 1 Q.B. 545; [1956] 2 W.L.R. 677; (1956) 100 S.J. 186; [1956] 1 All E.R. 613; (1956) 54 L.G.R. 212 65
Chrysler (U.K.) Ltd. *v.* McCarthy [1978] I.C.R. 939 .... 139, 146
Conway *v.* Rimmer [1968] A.C. 910; [1968] 2 W.L.R. 998; (1968) 112 S.J. 191; [1968] 1 All E.R. 874, H.L.; reversing [1967] 1 W.L.R. 1031 185
Crompton (A.) Ltd. *v.* Customs and Excise Commissioners (No. 2) [1974] A.C. 405; [1973] 3 W.L.R. 268; (1973) 117 S.J. 602; [1973] 2 All E.R. 1169, H.L.; affirming [1972] 2 Q.B. 102 180

D.H. Tools Co. *v.* Myers (1978) 1 D.S. Brief 138 155
Datsun Teesside Ltd. *v.* Stockton-on-Tees Borough Council, Case No. HS/25442/80; 28596/80; 28597/80 (C.O.I.T. No. 9/147/36) 150
Davie *v.* New Merton Board Mills Ltd. [1959] A.C. 604; [1959] 2 W.L.R. 331; (1959) 103 S.J. 177; [1959] 1 All E.R. 346; [1959] 2 Lloyd's Rep. 587, H.L.; affirming [1958] 1 Q.B. 210 92
Davis and Sons *v.* Environmental Health Department of Leeds City Council [1976] I.R.L.R. 282 138
Defrenne *v.* Sabena [1976] I.C.R. 547; [1976] E.C.R. 455; [1976] 2 C.M.L.R. 98 246
Dicker (George) & Sons *v.* Hilton, 9.2.79 (C.O.I.T. HS 2/102) 140, 149
Dickson *v.* Flack [1953] 2 Q.B. 464; [1953] 3 W.L.R. 571; (1953) 97 S.J. 586; [1953] 2 All E.R. 840; (1953) 51 L.G.R. 515; reversing [1953] 1 W.L.R. 196 111
Dorset (George) Ltd. *v.* Hill, 6.11.78 (C.O.I.T. HS 2/44) 153
Dowsett *v.* Ford Motor Company Ltd., 28.11.80 (C.O.I.T. 1975/235) 202
Drew *v.* St. Edmundsbury B.C. [1980] I.C.R. 513; [1980] I.R.L.R. 459 29
Dugdale *v.* Lovering (1875) L.R. 10 C.P. 196; (1875) 44 L.J.C.P. 197; (1875) 32 L.T. 155; (1875) 23 W.R. 391 158
Dunne *v.* North Western Gas Board [1964] 2 Q.B. 806; [1964] 2 W.L.R. 164; (1964) 107 S.J. 890; [1963] 3 All E.R. 916; (1963) 62 L.G.R. 197 230
Dutton *v.* Bognor Regis U.D.C. [1972] 1 Q.B. 373; [1972] 2 W.L.R. 299; (1971) 116 S.J. 16; [1972] 1 All E.R. 462; (1971) 70 L.G.R. 57; [1972] 1 Lloyd's Rep. 227, C.A.; affirming [1971] 2 All E.R. 1003 159
Dyson (R.A.) and Co. Ltd. *v.* Bentley, 26.11.79 (HS 2/151) 151

Ebbs *v.* Whitson (James) & Co. Ltd. [1952] 2 Q.B. 877; [1952] 1 T.L.R. 1428; (1952) 96 S.J. 375; [1952] 2 All E.R. 192; (1952) 50 L.G.R. 563 76
Edwards *v.* National Coal Board [1949] 1 K.B. 704; (1949) 65 T.L.R. 430; (1949) 93 S.J. 337; [1949] 1 All E.R. 743 ..... 63, 66
E.C. Commission *v.* Italy [1973] E.C.R. 101; [1973] C.M.L.R. 439 246

Fantarrow *v.* Leworthy (HS 21905/79) (unreported) 69
Fatima Development Corporation Ltd., *Re* [1974] 2 Ch. 271 158
Fife Tile Distributors Ltd. *v.* Mitchell, 14.10.77 (HS/6/77) 148
Finch *v.* Telegraph Construction and Maintenance Co. [1949] W.N. 57; (1949) 65 T.L.R. 153; (1949) 93 S.J. 219; [1949] 1 All E.R. 452; (1949) 47 L.G.R. 710 106

Galashiels Gas Co. *v.* O'Donnell [1949] A.C. 275; [1949] L.J.R. 540; (1949) 65 T.L.R. 76; (1949) 93 S.J. 71; [1949] 1 All E.R. 319; (1949) 47 L.G.R. 213, H.L.; affirming 1948 S.C. 191 9
Gannon *v.* Firth Ltd. [1977] I.T.R. 29; [1976] I.R.L.R. 415 26
Gateway Coal Co. *v.* U.M.W. 414 U.S. 368 (1974) 30
Geddis *v.* Proprietors of the Bann Reservoir (1878) 3 App.Cas. 430 ..... 230, 231
General Cleaning Contractors Ltd. *v.* Christmas [1953] A.C. 180; [1953] 2 W.L.R. 6; (1953) 97 S.J. 7; [1952] 2 All E.R. 1110; (1952) 51 L.G.R. 109, H.L.; affirming [1952] 1 K.B. 141 73
Ginty *v.* Belmont Building Supplies Ltd. [1959] 1 All E.R. 414 105
Grad. *v.* Finanzamt Traunstein [1970] E.C.R. 825; [1971] C.M.L.R. 1 246
Graham *v.* Holmes (1978) (HS 38151/77), Ashford Tribunal 138
Gregory *v.* Ford [1951] 1 All E.R. 121 172
Griffiths *v.* Dudley (Earl of) (1882) 9 Q.B.D. 357; (1882) 51 L.J.Q.B. 543; (1882) 47 L.T. 10; (1882) 46 J.P. 711; (1882) 30 W.R. 797 2
Groves *v.* Wimborne (Lord) [1898] 2 Q.B. 402; (1898) 67 L.J.Q.B. 862; (1898) 79 L.T. 284; (1898) 14 T.L.R. 493; (1898) 42 S.J. 633; (1898) 47 W.R. 87 173

Haigh *v.* Ireland (Charles W.) Ltd. [1974] 1 W.L.R. 43; (1974) 117 S.J. 939; [1973] 3 All E.R. 1137; (1973) 15 K.I.R. 283, H.L.; affirming 1973 S.L.T. 142 72
Harrison (T.C.) (Newcastle-under-Lyme) Ltd. *v.* Ramsey [1976] I.R.L.R. 135 ..... 148, 149
Haverson & Sons Ltd. *v.* Winger, 10.9.80 (HS 19536/80/F), South London Tribunal 139
Hixon *v.* Whitehead, 22.7.80 (C.O.I.T. HS 2/218) 154
Hollington *v.* Hewthorn [1943] K.B. 587; (1943) 112 L.J.K.B. 463; (1943) 169 L.T. 21; (1943) 59 T.L.R. 321; (1943) 87 S.J. 247; [1943] 2 All E.R. 35 175
Home Office *v.* Dorset Yacht Co. Ltd. [1970] A.C. 1004; [1970] 2 W.L.R. 1140; (1970) 114 S.J. 375; [1970] 2 All E.R. 294; [1970] 1 Lloyd's Rep. 453, H.L.; affirming [1969] 2 Q.B. 412, C.A.; affirming (1969) 113 S.J. 57 159
Hoover *v.* Mallon (1978) I.D.S. Brief 131 ..... 73, 153
Hudson *v.* Ridge Manufacturing Co. Ltd. [1957] 2 Q.B. 348; [1957] 2 W.L.R. 948; (1957) 101 S.J. 409; [1957] 2 All E.R. 229 103

Imperial Chemical Industries Ltd. *v.* Shatwell [1965] A.C. 656; [1964] 3 W.L.R. 329; (1964) 108 S.J. 578; [1964] 2 All E.R. 999, H.L.; reversing [1963] C.L.Y. 2377 101
Industrial Union Department, A.F.L.–C.I.O. *v.* American Petroleum Institute 65 L.Ed. 2d. 1010 (1980) 64, 121
—— *v.* Hodgson 499 F. 2d. 467 64

Jenkins *v.* Allied Ironfounders [1970] 1 W.L.R. 304; (1970) 114 S.J. 71; [1969] 3 All E.R. 1609; (1969) 8 K.I.R. 801 63

Kaukul *v.* Anglo-Soviet Shipping Co. (1931) 41 Lloyd's L.Rep. 90 23

Latimer *v.* A.E.C. [1953] A.C. 643; [1953] 3 W.L.R. 259; (1953) 117 J.P. 387; (1953) 97 S.J. 486; [1953] 2 All E.R. 449; (1953) 51 L.G.R. 457, H.L.; affirming [1952] 2 Q.B. 701, C.A.; varying [1952] 1 T.L.R. 507 81
Lewis *v.* Sweet, 18.2.81 (C.O.I.T. HS 3/19) 149

Lindsay *v.* Dunlop Ltd. [1980] I.R.L.R. 93 26
Lister *v.* Romford Ice and Cold Storage Co. [1957] A.C. 555; [1957] 2 W.L.R. 158; (1957) 121 J.P. 98; (1957) 101 S.J. 106; [1957] 1 All E.R. 125; [1956] 2 Lloyd's Rep. 505, H.L.; affirming [1956] 2 Q.B. 180 157

MacDermott *v.* Booth (1977) (unreported), Swansea Industrial Tribunal 140
McNeil *v.* Wane (1978) (unreported) 144
Manchester Corporation *v.* Farnworth [1930] A.C. 171; (1930) 99 L.J.K.B. 83; (1930) 46 T.L.R. 85; (1930) 73 S.J. 818; (1930) 94 J.P. 62; [1929] All E.R.Rep. 90 91
Manders Property Ltd. *v.* Johnson, 23.7.80 (C.O.I.T. HS 2/216) 138, 149
Marsh *v.* Judge International Housewares Ltd. (1977) (C.O.I.T. 571/57) 26, 103
Marshall *v.* Gotham Co. [1954] A.C. 360; [1954] 2 W.L.R. 812; (1954) 98 S.J. 268; [1954] 1 All E.R. 937, H.L.; affirming [1953] 1 Q.B. 167 .... 63, 65, 67
Matthews *v.* Kuwait Bechtel Corporation [1959] 2 Q.B. 57; [1959] 2 W.L.R. 702; (1959) 103 S.J. 393; [1959] 2 All E.R. 345 23
Mayhew *v.* Anderson [1978] I.R.L.R. 101 103
Miles *v.* Forest Rock Granite Co. (Gloucestershire) Ltd. (1918) 34 T.L.R. 500; (1918) 62 S.J. 634 230
Miller *v.* Boothman (William) and Sons Ltd. [1944] K.B. 337; (1944) 113 L.J.K.B. 206; (1944) 170 L.T. 187; (1944) 60 T.L.R. 218; [1944] 1 All E.R. 333 111
—— *v.* Ministry of Pensions [1947] W.N. 241; [1948] L.J.R. 203; (1948) 177 L.T. 536; (1948) 63 T.L.R. 474; (1948) 91 S.J. 484; [1947] 2 All E.R. 372 174
Morris *v.* Wilkins (1979) (unreported) 144
Murray *v.* Gadbury Q.B.D. (D.C.) (HS 13127/77) 147
—— *v.* Schwachman Ltd. [1938] 1 K.B. 130; (1937) 106 L.J.K.B. 354; (1937) 156 L.T. 407; (1937) 53 T.L.R. 458; (1937) 81 S.J. 294; [1937] 2 All E.R. 68; (1937) 30 B.W.C.C. 466 106

N.A.A.F.I. *v.* Portsmouth City Council, 8.1.79 (C.O.I.T. HS 2/54) 149
N.L.R.B. *v.* Washington Aluminium Co. 370 U.S. 9 (1962) 29
Nicholls *v.* Austin (Leyton) Ltd. [1946] A.C. 493; (1946) 115 L.J.K.B. 329; (1946) 175 L.T. 5; (1946) 62 T.L.R. 320; (1946) 90 S.J. 628; [1946] 2 All E.R. 92; (1946) 44 L.G.R. 287 6
Nico Manufacturing Co. Ltd. *v.* Hendry [1975] I.R.L.R. 224 153
Nimmo *v.* Cowan (Alexander) and Sons Ltd. [1968] A.C. 107; [1967] 3 W.L.R. 1169; (1967) 111 S.J. 668; [1967] 3 All E.R. 187; [1967] 3 K.I.R. 277, H.L.; reversing 1966 S.L.T. 266 .... 9, 174
Norris *v.* Syndic Manufacturing Co. [1952] 2 Q.B. 135; [1952] 1 T.L.R. 858; [1952] 1 All E.R. 935 .... 71, 106
Northampton B.C. *v.* Farthingstone Silos Ltd. (1981) (unreported), Northampton Crown Court 88

Ogden *v.* London Electric Railway Company (1933) 149 L.T. 476; (1933) 49 T.L.R. 542; [1933] All E.R.Rep. 896 180
Osborne *v.* Taylor (Bill) of Huyton [1982] I.R.L.R. 17 83
Otterburn Mill Ltd. *v.* Bulman [1975] I.R.L.R. 223 .... 147, 154
Ottoman Bank *v.* Chakarian [1930] A.C. 277; (1930) 99 L.J.P.C. 97; (1930) 142 L.T. 465 .... 23, 24
Oxley (Graham) Tool Steels Ltd. *v.* Firth [1980] I.R.L.R. 135 .... 24, 25, 139

Pagano *v.* HGS [1976] I.R.L.R. 9 24
Page *v.* Freight Hire (Tank Haulage) Ltd. [1981] I.C.R. 299; [1981] 1 All E.R. 394; [1981] I.R.L.R. 13 77
Politi *v.* Italian Ministry of Finance [1971] E.C.R. 1039 246
Porter *v.* Bandridge [1978] I.C.R. 943; [1978] 1 W.L.R. 1145; (1978) 122 S.J. 592; (1978) 13 I.T.R. 340; [1978] I.R.L.R. 271 69
Porthole Ltd. *v.* Brown, 7.3.80 (C.O.I.T. HS 2/186) 151
Powell *v.* Phillips (1972) 116 S.J. 713; [1972] 3 All E.R. 864; [1973] R.T.R. 19 118
Powley *v.* British Siddeley Engines Ltd. [1966] 1 W.L.R. 729; (1966) 110 S.J. 369; [1965] 3 All E.R. 612 65
Preedy (Alfred) & Sons Ltd. *v.* Owens, 1.5.79 (C.O.I.T. HS 2/102) 140

Price *v.* Claudgen [1967] 1 W.L.R. 48; (1967) 111 S.J. 176; [1967] 1 All E.R. 695; (1967) 2 K.I.R. 127 6

Qualcast (Wolverhampton) Ltd. *v.* Haynes [1959] A.C. 743; [1959] 2 W.L.R. 510; (1959) 103 S.J. 310; [1959] 2 All E.R. 38, H.L.; reversing [1958] 1 W.L.R. 225 67

R. *v.* Caldwell [1981] 2 W.L.R. 509; (1981) 125 S.J. 239; [1981] 1 All E.R. 961; (1981) 73 Cr.App.R. 13, H.L.; affirming (1980) 71 Cr.App.R. 237 105
—— *v.* Dudley and Stephens (1884) 14 Q.B.D. 273; (1884) 54 L.J.M.C. 32; (1884) 52 L.T. 107; (1884) 1 T.L.R. 118; (1884) 49 J.P. 69; (1884) 33 W.R. 347 21
—— *v.* Lawrence [1981] 2 W.L.R. 524; (1981) 125 S.J. 241; [1981] 1 All E.R. 974; (1981) 73 Cr.App.R. 1; [1981] R.T.R. 217, H.L.; reversing (1980) 71 Cr.App.R. 291 105
—— *v.* Miller (Robert) (Contractors) Ltd. [1970] 2 Q.B. 54; [1970] 2 W.L.R. 541; [1970] 1 All E.R. 577; [1970] R.T.R. 147; (1970) 54 Cr.App.R. 158, C.A.; affirming [1969] 3 All E.R. 247 170
—— *v.* Swan Hunter Shipbuilders Ltd. [1981] I.C.R. 831; [1982] 1 All E.R. 264; [1981] I.R.L.R. 403; [1981] 2 Lloyd's Rep. 605; [1981] Crim.L.R. 833........73, 75, 86, 87, 167
Raphael, dec'd, *Re*; Raphael *v.* D'Antin [1973] 1 W.L.R. 998; (1973) 117 S.J. 566; [1973] 3 All E.R. 19 175
Read *v.* Lyons [1947] A.C. 156; [1947] L.J.R. 39; (1947) 175 L.T. 413; (1947) 62 T.L.R. 646; (1947) 91 S.J. 54; [1946] 2 All E.R. 471 ....229, 230
Rees *v.* Cambrian Wagon Works Ltd. (1946) 175 L.T. 220; (1946) 62 T.L.R. 512; (1946) 90 S.J. 405 73
Rice *v.* Connolly [1966] 2 Q.B. 414; [1966] 3 W.L.R. 17; (1966) 130 J.P. 322; (1966) 110 S.J. 371; [1966] 2 All E.R. 649 135
Roadline (U.K.) Ltd. *v.* Mainwaring (1977) (unreported), Swansea Industrial Tribunal 140
Russell *v.* Kelly, 15.6.77 (HS 2/77—Scotland) 151
Rutili *v.* French Minister of the Interior [1975] E.C.R. 1219; [1976] I.C.M.L.R. 140 ... 246
Rylands *v.* Fletcher (1868) L.R. 3 H.L. 330, H.L.; affirming (1866) L.R. 1 Ex. 265; (1866) 4 H. & C. 263; (1866) 35 L.J.Ex. 154; (1866) 14 L.T. 523; (1866) 30 J.P. 436; (1866) 12 Jur.N.S. 603; (1866) 14 W.R. 799........229, 230, 231

St. Anne's Board Mills Co. Ltd. *v.* Brien [1973] I.C.R. 444; [1973] I.T.R. 463; [1973] I.R.L.R. 309 27
Salmon *v.* Cooper, 30.4.80 (C.O.I.T. HS 2/199) 147
Scholefield *v.* Schunck (1855) 19 J.P. 84; (1855) 24 L.T.O.S. 253 91
Science Research Council *v.* Nassé [1980] A.C. 1028; [1979] I.C.R. 921; [1979] 3 W.L.R. 762; (1979) 123 S.J. 768; [1979] 3 All E.R. 673; [1979] I.R.L.R. 465, H.L.; affirming [1979] Q.B. 144 186
Shipbreaking Q. Ltd. *v.* Tonge, 17.1.80 (HS 2/162) 147
Simmonds *v.* U.S.A. (1976) (C.O.I.T. 561/214) 103
Siveyer *v.* Randall (1978) (unreported) 143
Skinner *v.* McGregor (John G.) (Contractors) Ltd. 1977 S.L.T. 83 133
Smith *v.* Baker & Sons Ltd. [1891] A.C. 325; (1891) 60 L.J.Q.B. 683; (1891) 65 L.T. 467; (1891) 7 T.L.R. 697; (1891) 55 J.P. 660; (1891) 40 W.R. 392 2
—— *v.* Central Asbestos Co. *See* Central Asbestos Co. *v.* Dodd.
—— *v.* National Coal Board [1967] 1 W.L.R. 871; (1967) 111 S.J. 455; [1967] 2 All E.R. 593; (1967) 3 K.I.R. 1, H.L.; reversing [1966] 1 W.L.R. 682 65
South Surbiton Cooperative Society *v.* Wilcox [1975] I.R.L.R. 293 147
Squibb *v.* United Kingdom Staff Association [1979] I.C.R. 235; [1979] 1 W.L.R. 523; (1978) 123 S.J. 352; [1979] 2 All E.R. 452; [1979] I.R.L.R. 75, C.A.; reversing [1978] I.C.R. 115 193
Stephensons (Crane Hire) Ltd. *v.* Gordon (H.M. Inspector of Health and Safety), 17.8.77 (HS/8868/77) 152
Summers (John) & Sons Ltd. *v.* Frost [1955] A.C. 740; [1955] 2 W.L.R. 825; (1955) 99 S.J. 257; [1955] 1 All E.R. 870; (1955) 53 L.G.R. 329, H.L.; affirming [1954] 2 Q.B. 21 7

T.W. Enamellers Ltd. *v.* Chapman, 19.5.77 (unreported) 149
Taylor *v.* Alidair Ltd. [1978] I.C.R. 445, C.A.; affirming [1977] I.C.R. 446; (1977) 121 S.J. 758; (1976) 12 I.T.R. 21; [1976] I.R.L.R. 420 27
—— *v.* Bowater Flexible Packaging Co., 9.5.73 (unreported) 103
—— *v.* Rover Co. Ltd. [1966] 1 W.L.R. 1491; [1966] 2 All E.R. 181........ 71, 92
Tesco Supermarkets *v.* Nattrass [1972] A.C. 153; [1971] 2 W.L.R. 1166; (1971) 115 S.J. 285; [1971] 2 All E.R. 127; (1971) 69 L.G.R. 403, H.L.; reversing [1971] 1 Q.B. 133 170
Thurogood *v.* Van Den Berghs and Jurgens Ltd. [1951] 2 K.B. 537; [1951] 1 T.L.R. 557; (1951) 115 J.P. 237; (1951) 95 S.J. 317; (1951) L.G.R. 504; [1951] 1 All E.R. 682 72
Trott *v.* Smith (W.E.) [1957] 1 W.L.R. 1154; (1957) 101 S.J. 885; [1957] 3 All E.R. 500; (1957) 56 L.G.R. 20 65
Tyrematic Ltd. *v.* Cottle, 21.2.80 (C.O.I.T. HS 2/185)........ 139, 147

Union of Shop, Distributive and Allied Workers *v.* Sketchley Ltd. [1981] I.C.R. 644; [1981] I.R.L.R. 291 193

Vacwell Engineering Ltd. *v.* B.D.H. Chemicals Ltd. [1971] 1 Q.B. 111; [1970] 3 W.L.R. 67; (1970) 114 S.J. 472; [1970] 3 All E.R. 553n., C.A.; varying [1971] 1 Q.B. 88 .... 79, 98
Van Duyn *v.* Home Office [1975] Ch. 358; [1975] 2 W.L.R. 760; (1974) 119 S.J. 302; [1975] 3 All E.R. 190; [1974] E.C.R. 1337; [1975] 1 C.M.L.R. 1 246
Verbonde van Nederlandse Ondernemingen *v.* Inspecteur der Invoer rechten en Accijen [1977] E.C.R. 113; [1977] 1 C.M.L.R. 413 246
Vinyl Compositions Ltd. *v.* Barnard, 12.8.80 (C.O.I.T. HS 3/15) 148

Warner Group of Companies *v.* Smith, 18.10.79 (HS 2/134) 151
Watling *v.* Bird (William) and Son (Contractors) Ltd. (1976) 11 I.T.R. 70 146
Waugh *v.* British Railways Board [1980] A.C. 521; [1979] 3 W.L.R. 150; (1979) 123 S.J. 506; [1979] 2 All E.R. 1169; [1979] I.R.L.R. 364, H.L.; reversing (1978) 122 S.J. 730 180
Wearing *v.* Pirelli [1977] I.C.R. 90; [1977] 1 W.L.R. 48; [1977] 1 All E.R. 339; [1979] I.R.L.R. 36 6
Westall Richardson Ltd. *v.* Roulson [1954] 1 W.L.R. 905; (1954) 98 S.J. 423; [1954] 2 All E.R. 448 89
Westergren *v.* Cox (1978) (unreported) 144
Western Excavating (ECC) Ltd. *v.* Sharp [1978] Q.B. 761; [1978] I.C.R. 221; [1978] 2 W.L.R. 344; (1977) 121 S.J. 814; [1978] 1 All E.R. 713; (1978) 13 I.T.R. 132, C.A.; reversing [1978] I.R.L.R. 27 24
Whirlpool Corporation *v.* Marshall 445 U.S. 1 (1980) 30
White *v.* Pressed Steel Fisher [1980] I.R.L.R. 176 204
Wilcox *v.* Humphreys & Glasgow Ltd. [1976] I.C.R. 306, C.A.; affirming [1975] I.C.R. 333; [1975] I.T.R. 103; [1975] I.R.L.R. 211 26
Wilkinson *v.* Franks (1978) 1 D.S. Brief 143........ 153, 156
Wilson *v.* Tyneside Window Cleaning Co. [1958] 2 Q.B. 110; [1958] 2 W.L.R. 900; (1958) 102 S.J. 380; [1958] 2 All E.R. 265 74
Wilsons and Clyde Coal Co. *v.* English [1938] A.C. 79; (1938) 108 L.J.P.C. 117; (1938) 157 L.T. 406; (1938) 53 T.L.R. 944; (1938) 81 S.J. 700; [1937] 3 All E.R. 628 73
Winter *v.* Cardiff R.D.C. [1950] W.N. 193; (1950) 114 J.P. 234; [1950] 1 All E.R. 819; (1950) 49 L.G.R. 1 73
Wright *v.* Ford Motor Co. Ltd. [1967] 1 Q.B. 230; [1966] 3 W.L.R. 112; (1966) 110 S.J. 370; [1966] 2 All E.R. 518........ 104, 106

TABLE OF STATUTES

1802 Factory Act (42 Geo. 3, c. 73) 3, 161
1833 Labour of Children etc. in Factories (3 & 4 Will. 4, c. 103) 50
1844 Factory Act (7 & 8 Vict. c. 15) 50, 91, 161
1872 Coal Mines Regulation Act (35 & 36 Vict. c. 76)—
General Rule 30 191
1874 Hosiery Manufacture (Wages) Act (37 & 38 Vict. c. 48)—
s. 2 107
1875 Explosives Act (38 & 39 Vict. c. 17) 177, 223
Public Health Act (38 & 39 Vict. c. 55)—
s. 265 159
1878 Factory and Workshop Act (41 & 42 Vict. c. 16) 3
s. 3(4) 173
1880 Employers' Liability Act (43 & 44 Vict. c. 42) 1, 2
1896 Truck Act (59 & 60 Vict. c. 44)—
s. 3 107
1906 Alkali etc. Works Regulation Act (6 Edw. 7, c. 14)—
s. 27(1) 90, 91
1911 Coal Mines Act (1 & 2 Geo. 5, c. 50) 31
s. 16 191
s. 102(8) 63
1919 Industrial Courts Act (9 & 10 Geo 5, c. 69)—
s. 8 89
1923 Explosives Act (13 & 14 Geo. 5, c. 17) 223
1928 Petroleum (Consolidation) Act (18 & 19 Geo. 5, c. 32) 221
s. 5 221
1933 Children and Young Persons Act (23 & 24 Geo. 5, c. 12) 161
1936 Public Health Act (26 Geo. 5 & Edw. 8, c. 49) 159
1945 Law Reform (Contributory Negligence) Act (8 & 9 Geo. 6, c. 28) 2
1947 Statistics of Trade Act (10 & 11 Geo. 6, c. 39)—
s. 9 181
Crown Proceedings Act (10 & 11 Geo. 6, c. 44) 158
s. 10(7) 158
1948 Law Reform (Personal Injuries) Act (11 & 12 Geo. 6, c. 41)—
s. 1(1) 2
1949 Civil Aviation Act (12, 13 & 14 Geo. 6, c. 67) 234
1950 Shops Act (14 Geo. 6, c. 28) 220
1951 Fireworks Act (14 & 15 Geo. 6, c. 58) 223
1954 Mines and Quarries Act (2 & 3 Eliz. 2, c. 70) 4, 81, 104, 171
s. 123 191
ss. 135–137 79
s. 180 192
1956 Agriculture (Safety, Health and Welfare Provisions) Act (4 & 5 Eliz. 2, c. 49) 4, 5
1961 Factories Act (9 & 10 Eliz. 2, c. 34) 4, 56, 72, 76, 82, 108, 139, 144, 165, 166, 174
s. 1 148
(3) 143
s. 2 9
s. 3 139
(1) 24
s. 4 9
s. 13(1) 66
s. 14 108
(1) 6, 66, 106, 111
s. 17 .... 92, 93
ss. 28, 29 81
s. 29 .... 9, 175
(1) 81
s. 36 .... 140, 149
s. 58 9
s. 63(1) 26
s. 69 55
s. 76 111
s. 116 5
s. 119 105
s. 121 144
s. 136 107
s. 138 79
s. 143 104
s. 175 82
s. 176(1) 57
1963 Offices, Shops and Railway Premises Act (c. 41) 4, 56, 57, 60, 82, 126, 144, 165
s. 6 139
(1) 149
(3)(*b*) 149
s. 9 138
s. 16 .... 81, 139
s. 27 104

1963 Offices, Shops and Railway Premises Act—*cont*
s. 50 ... 79
s. 52(5) ... 57
s. 62 ... 143
1964 Industrial Training Act (c. 16) ... 155
Continental Shelf Act (c. 29) ... 37
Police Act (c. 48)—
s. 62 ... 57
1965 Gas Act (c. 36) ... 232, 234
Nuclear Installations Act (c. 57) ... 232, 233, 234
s. 1 ... 232
s. 2 ... 232
ss. 3–5 ... 232
s. 7 ... 232
s. 11 ... 232
s. 12 ... 172
s. 13(4)(*a*) ... 232
s. 16 ... 233
s. 22 ... 233
s. 24 ... 232
Redundancy Payments Act (c. 62) ... 114
1967 Parliamentary Commissioner Act (c. 13)—
Sched. 1 ... 153
Companies Act (c. 81) ... 186
1968 Civil Evidence Act (c. 64)—
s. 11 ... 176
(1) ... 175
(2) ... 175
Medicines Act (c. 67)—
s. 75 ... 220
Transport Act (c. 73) ... 46
1969 Employers' Liability (Defective Equipment Act) (c. 37) ... 92
Post Office Act (c. 48)—
s. 11 ... 193
Employers' Liability (Compulsory Insurance) Act (c. 57) ... 2, 136
1970 Administration of Justice Act (c. 31)—
s. 31 ... 185
s. 32 ... 185
(1) ... 185
Chronically Sick and Disabled Persons Act (c. 44)—
ss. 4 and 8 ... 82
1971 Fire Precautions Act (c. 40) ... 60
s. 17 ... 60
s. 43(1) ... 57
Mineral Workings (Offshore Installations) Act (c. 61) ... 38
Tribunals and Inquiries Act (c. 62) ... 157
1971 Industrial Relations Act (c. 72) ... 186
s. 106 ... 69
1972 Road Traffic Act (c. 20) ... 118
s. 37 ... 118
Employment Medical Advisory Services Act (c. 28) ... 4, 50
Contracts of Employment Act (c. 53) ... 114
Local Government Act (c. 70)—
s. 250 ... 60
1974 Local Government Act (c. 7) ... 153
Health and Safety at Work etc. Act (c. 37) ... 1, 5, 7, 8, 9, 10, 11, 12, 14, 17, 28, 31, 33, 35, 37, 38, 39, 47, 49, 50, 54, 55, 56, 58, 60, 62, 90, 91, 110, 114, 115, 119, 120, 123, 124, 126, 139, 143, 144, 164, 165, 166, 167, 169, 171, 174, 175, 176, 192, 206, 212, 234, 256
s. 1(1) ... 61
(*d*) ... 90
(2) ... 61
s. 2 ... 65, 74, 77, 89, 93, 108, 120, 140, 165, 210
(1) ... 66, 70, 75, 86, 103, 149, 152, 170
(2) ... 71, 72, 144
(*a*) ... 72, 73, 74, 75
(*b*) ... 76
(*c*) ... 28, 75, 78, 79, 80, 84, 95, 204
(*d*) ... 76, 80, 81
(*e*) ... 82
(3) ... 12, 78, 83, 212, 213, 215
(4) ... 84, 187, 191, 194, 198, 201, 203
(4), (6) and (7) ... 78
(4)–(7) ... 84
(5) ... 84, 191
(6) ... 194, 198
(7) ... 205, 206
ss. 2 and 4 ... 93
ss. 2–7 ... 102, 117, 162, 166, 172
ss. 2–9 ... 62, 73, 111, 112, 164
s. 3 ... 75, 85, 86, 89
(1) ... 86, 87
(1) and (2) ... 152
(2) ... 87
(3) ... 79, 87, 116
ss. 3–6 ... 165
s. 4 ... 57, 86, 87, 88, 89, 138
(1)(*b*) ... 89
(2) ... 82, 144
s. 5 ... 90
(1) ... 90
(2) ... 90
(3) ... 90

1975 Employment Protection Act—*cont.*
ss. 11–16 ... 193
s. 17 ... 188
ss. 17–21 ... 186
s. 88(3) ... 193
s. 99 ... 186
s. 116 ... 47
ss. 116, 125(3) ... 191
Sched. 15, para. 2 ... 191
Sched. 18 ... 191
Petroleum Submarine Pipe-Lines Act (c. 74) ... 38
1976 Chronically Sick and Disabled Persons (Amendment) Act (c. 49)—
s. 2 ... 82
Local Government (Miscellaneous Provisions) Act (c. 57)—
ss. 27 and 39 ... 159
1977 Criminal Law Act (c. 45) ... 163
ss. 14, 64(1) ... 163
ss. 19–26 ... 163
s. 21(3)(*b*) ... 163
s. 60(1) ... 173
s. 64(1) ... 163
1977 Criminal Law Act—*cont.*
Sched. 1 ... 164
Sched. 6 ... 164
Sched. 12 ... 164
1978 Consumer Safety Act (c. 38) 92, 93
s. 1 ... 99
(4) ... 93
Employment Protection (Consolidation) Act (c. 44) 69, 114
s. 1(5) ... 28
s. 57(3) ... 25, 26
s. 58(1)(*b*) ... 28, 29
1979 Wages Councils Act (c. 12)—
s. 28 ... 89
Sched. 2, para. 1(2) ... 198
1980 Employment Act (c. 42) ... 193
s. 19 ... 193
Magistrates' Court Act (c. 43)—
s. 101 ... 96
s. 111(2) ... 168
s. 127(1) ... 168
1981 Disabled Persons Act (c. 43)—
s. 6 ... 82
Employment and Training Act (c. 57)—
s. 9 ... 181

TABLE OF STATUTORY INSTRUMENTS

1949 Packing of Explosive for Conveyance Rules (S.I. 1949 No. 798) 223

1960 Factories (Cleanliness of Walls and Ceilings) Order (S.I. 1960 No. 1794) 150

1969 Asbestos Regulations (S.I. 1969 No. 690) 77, 167

1970 Abrasive Wheels Regulations (S.I. 1970 No. 535) 7
- Regs. 2(2) and 3(2) 111

1972 Industrial Tribunals (Industrial Relations etc.) Regulations (S.I. 1972 No. 38)—
- Sched., Rule 2(1) 69

Highly Flammable Liquids and Liquefied Petroleum Gases Regulations (S.I. 1972 No. 917) 221
- Reg. 6 221

1974 Woodworking Machines Regulations (S.I. 1974 No. 903) 147

Offshore Installations (Diving Operations) Regulations (S.I. 1974 No. 1229) 51

Health and Safety at Work Act (Commencement No. 1) Order (1974 No. 1439) 46, 256

Radioactive Substances (Carriage by Road) (Great Britain) Regulations (S.I. 1974 No. 1735) 223

Anthrax Prevention Act 1919 (Repeals and Modifications) Regulations (S.I. 1974 No. 1775) 256

Factories Act 1961 (Enforcement of Section 135) Regulations (S.I. 1974 No. 1776) 256

Docks and Harbours Act 1966 (Modification) Regulations (S.I. 1974 No. 1820) 256

Radioactive Substance Act 1948 (Modification) Regulations (S.I. 1974 No. 1821) 256

Hydrogen Cyanide (Fumigation) Act 1937 (Repeals and Modifications) Regulations (S.I. 1974 No. 1840) 256

1974 Celluloid and Cinematograph Film Act 1922 (Repeals and Modifications) Regulations (S.I. 1974 No. 1841) 256

Explosives Acts 1875 and 1923 etc. (Repeals and Modifications) Regulations (S.I. 1974 No. 1885) 256

Boiler Explosions Acts 1882 and 1890 (Repeals and Modifications) Regulations (S.I. 1974 No. 1886) 256

Truck Acts 1831 to 1896 (Enforcement) Regulations (S.I. 1974 No. 1887) 256

Industrial Tribunals (Improvement and Prohibition Notices Appeals) Regulations (S.I. 1974 No. 1925) 155, 256
- Reg. 3 68
- Sched., Rule 2(2) 68
  - Rule 12 157
  - Rule 13 156
  - Rule 14(3) 69

Industrial Tribunals (Improvement and Prohibition Notices Appeals) (Scotland) Regulations (S.I. 1974 No. 1926) 155, 256

Factories Act 1961 etc. (Repeals and Modifications) Regulations (S.I. 1974 No. 1941) 256

Petroleum (Regulation) Acts 1928 and 1936 (Repeals and Modifications) Regulations (S.I. 1974 No. 1942) 256

Offices, Shops and Railway Premises Act 1963 (Repeals and Modifications) Regulations (S.I. 1974 No. 1943) 256

Pipe-lines Act 1962 (Repeals and Modifications) Regulations (S.I. 1974 No. 1986) 256

Coal Industry Nationalisation Act 1946 (Repeals) Regulations (S.I. 1974 No. 2011) 256

Ministry of Fuel and Power Act 1945 (Repeal) Regulations (S.I. 1974 No. 2012) 256

1974 Mines and Quarries Acts 1954 to 1971 (Repeals and Modifications) Regulations (S.I. 1974 No. 2013) 256
Health and Safety Licensing Appeals (Hearings Procedure) Rules (S.I. 1974 No. 2040) 256
Nuclear Installations Act 1965 etc. (Repeals and Modifications) Regulations (S.I. 1974 No. 2056) 257
Health and Safety Licensing Appeals (Hearings Procedure) (Scotland) Rules (S.I. 1974 No. 2068) 257
Explosives Acts 1875 and 1923 etc. (Repeals and Modifications) (Amendment) Regulations (S.I. 1974 No. 2166) 257
Clean Air Enactments (Repeals and Modifications) Regulations (S.I. 1974 No. 2170) 257
1975 Agriculture (Poisonous Substances) Act 1952 (Repeals and Modifications) Regulations (S.I. 1975 No. 45) 257
Agriculture (Safety, Health and Welfare Provisions) Act 1956 (Repeals and Modifications) Regulations (S.I. 1975 No. 46) 257
Merchant Shipping (Diving Operations) Regulations (S.I. 1975 No. 116) 51
Health and Safety (Agriculture) (Poisonous Substances) Regulations (S.I. 1975 No. 282) 257
Protection of Eyes (Amendment) Regulations (S.I. 1975 No. 303) 257
Health and Safety Inquiries (Procedure) Regulations (S.I. 1975 No. 335).........45, 257
Offices, Shops and Railway Premises Act 1963 (Repeals) Regulations (S.I. 1975 No. 1011) 257
Factories Act 1961 (Repeals) Regulations (S.I. 1975 No. 1012) 257
Mines and Quarries Acts 1954 to 1971 (Repeals and Modifications) Regulations (S.I. 1975 No. 1102) 257
Coal Mines (Respirable Dust) Regulations (S.I. 1975 No. 1433) 257
1975 Agricultural or Forestry Tractors and Tractor Components (Type Approval) Regulations (S.I. 1975 No. 1475) 270
Employers' Health and Safety Policy Statements (Exception) Regulations (S.I. 1975 No. 1584)................83, 212, 257
Conveyance of Explosives by Road (Special Case) Regulations (S.I. 1975 No. 1621) 257
Baking and Sausage Making (New Year) Regulations (S.I. 1975 No. 1695) 257
1976 Coal Mines (Precautions against Inflammable Dust) Temporary Provisions Regulations (S.I. 1976 No. 881) 257
Submarines Pipelines (Diving Operations) Regulations (S.I. 1976 No. 923) 51
Operations at Unfenced Machinery (Amendment) Regulations (S.I. 1976 No. 955) 257
Health and Safety Inquiries (Procedure) (Amendment) Regulations (S.I. 1976 No. 1246) 257
Health and Safety (Agriculture) (Miscellaneous Repeals and Modifications) Regulations (S.I. 1976 No. 1247) 257
Building Regulations (S.I. 1976 No. 1676).............118, 138
Baking and Sausage Making (Christmas and New Year) Regulations (S.I. 1976 No. 1908) 257
Fire Certificates (Special Premises) Regulations (S.I. 1976 No. 2003) 257
Factories Act 1961 etc. (Repeals) Regulations (S.I. 1976 No. 2004) 257
Offices, Shops and Railway Premises Act 1963 (Repeals) Regulations (S.I. 1976 No. 2005) 257
Fire Precautions Act 1971 (Modifications) Regulations (S.I. 1976 No. 2007) 257
Mines and Quarries (Metrication) Regulations (S.I. 1976 No. 2063) 257

1977 Safety Representatives and Safety Committees Regulations (S.I. 1977 No. 500) 110, 120, 192, 194, 196, 203, 210, 218, 224, 257
Reg. 4 198
(1)(rider) 203
(*b*) 29
(2) 203
Regs. 5–7 201
Reg. 6 202
Reg. 7 203
(1) 187
(2) 187
(3) 188
Reg. 9 205
Reg. 11 197
Schedule 203
Health and Safety (Enforcing Authority) Regulations (S.I. 1977 No. 746) 53, 55, 56, 59, 149, 257
Reg. 3 .... 55, 57
Reg. 4 56
(2)(*a*) 59
(3) 57
Reg. 5 58
Reg. 6 58
(1) 58
(2) 59
Sched. 1 .... 55, 56, 57, 59
Coal Mines (Precautions against Inflammable Dust) Amendment Regulations (S.I. 1977 No. 913) 257
Explosives (Registration of Premises) Variation of Fees Regulations (S.I. 1977 No. 918) 258
Coal and Other Mines (Electricity) (Third Amendment) Regulations (S.I. 1977 No. 1205) 258
Health and Safety at Work etc. Act 1974 (Application outside Great Britain Order) (S.I. 1977 No. 1232) 37, 258
Acetylene (Exemption) Order (S.I. 1977 No. 1798) 258
Baking and Sausage Making (Christmas and New Year) Regulations (S.I. 1977 No. 1841) 258
1978 Poisons Rules (S.I. 1978 No. 1) 221, 223
Packaging and Labelling of Dangerous Substances Regulations (S.I. 1978 No. 209) 98, 219, 221, 258, 268
Reg. 2 220

1978 Packaging and Labelling of Dangerous Substances Regulations—*cont*
Reg. 5 .... 219, 220
Reg. 7(4) 221
Sched. 1 .... 219, 220
Sched. 2 220
Sched. 3 220
Sched. 4 220
Explosives (Licensing of Stores) Variation of Fees Regulations (S.I. 1978 No. 270) 258
Petroleum (Regulation) Acts 1928 and 1936 (Variation of Fees) Regulations (S.I. 1978 No. 635) 258
Health and Safety (Genetic Manipulation) Regulations (S.I. 1978 No. 752) 258
Reg. 3 71
Reg. 4 86
Coal Mines (Respirable Dust) Regulations (S.I. 1978 No. 807) 258
Medicines (Administration of Radioactive Substances) Regulations (S.I. 1978 No. 1006) 266
Health and Safety at Work (Northern Ireland) Order (S.I. 1978 No. 1039 (N.I. 9.)) 37
Factories (Standards of Lighting) Revocation Regulations (S.I. 1978 No. 1126) 258
Cosmetic Products Regulations (S.I. 1978 No. 1354) 221
Baking and Sausage Making (Christmas and New Year) Regulations (S.I. 1978 No. 1516) 258
Merchant Shipping (Dangerous Goods) Rules (S.I. 1978 No. 1543) 223
Coal and Other Mines (Metrication) Regulations (S.I. 1978 No. 1648) 258
Hazardous Substances (Labelling of Road Tankers) Regulations (S.I. 1978 No. 1702) 77, 258
Compressed Acetylene (Importation) Regulations (S.I. 1978 No. 1723) 258
Mines and Quarries Act 1954 (Modification) Regulations (S.I. 1978 No. 1951) 258

1979 Agricultural or Forestry Tractors and Tractor Components (Type Approval) Regulations (S.I. 1979 No. 221) 270
Mines (Precautions Against Inrushes) Regulations (S.I. 1979 No. 318) 258
Petroleum (Consolidation) Act 1928 (Enforcement) Regulations (S.I. 1979 No. 427) 258
Coal and Other Mines (Electric Lighting for Filming) Regulations (S.I. 1979 No. 1203) 258
Baking and Sausage Making (Christmas and New Year) Regulations (S.I. 1979 No. 1298) 258
Explosives Act 1875 (Exemptions) Regulations (S.I. 1979 No. 1378) 258
Health and Safety (Fees for Medical Examinations) Regulations (S.I. 1979 No. 1553) 258

1980 Health and Safety (Enforcing Authority) (Amendment) Regulations (S.I. 1980 No. 744) 59
Notification of Accidents and Dangerous Occurrences Regulations (S.I. 1980 No. 804) 110, 178, 202, 217, 218, 258
Reg. 2(1) 218
Sched. 2 217
Health and Safety (Leasing Arrangements) Regulations (S.I. 1980 No. 907) 99, 258
Coal and Other Mines (Fire and Rescue) (Amendment) Regulations (S.I. 1980 No. 942) 258
Agriculture (Tractor Cabs) (Amendment) Regulations (S.I. 1980 No. 1036) 258
Petroleum (Consolidation) Act 1928 (Conveyance by Road Regulations Exemptions) Regulations (S.I. 1980 No. 1100) 258
Mines and Quarries (Fees for Approvals) Regulations (S.I. 1980 No. 1233) 258
Control of Lead at Work Regulations (S.I. 1980 No. 1248) 77, 110, 224, 258, 266

1980 Celluloid and Cinematograph Film Act 1922 (Exemptions) Regulations (S.I. 1980 No. 1314) 259
Safety Signs Regulations (S.I. 1980 No. 1471) 119, 178, 259
Baking and Sausage Making (Christmas and New Year) Regulations (S.I. 1980 No. 1576) 259
Health and Safety (Animal Products) (Metrication) Regulations (S.I. 1980 No. 1690) 259
Control of Pollution (Special Waste) Regulations (S.I. 1980 No. 1709) 221
Health and Safety (Enforcing Authority) (Amendment) Regulations (S.I. 1980 No. 1744) 259

1981 Chemical Works (Metrication) Regulations (S.I. 1981 No. 16) 259
Mines and Quarries (Fees for Approvals) (Amendment) Regulations (S.I. 1981 No. 270) 259
Health and Safety (Fees for Medical Examinations) Regulations (S.I. 1981 No. 334) 259
Diving Operations at Work Regulations (S.I. 1981 No. 399) 38, 110, 259
Reg. 4 108
Agricultural or Forestry Tractors and Tractor Components (Type Approval) (Amendment) Regulations (S.I. 1981 No. 669) 270
Packaging and Labelling of Dangerous Substances (Amendment) Regulations (S.I. 1981 No. 792) 77, 98, 219, 268
Reg. 2(b) 219
Health and Safety (First Aid) Regulations (S.I. 1981 No. 917) 224, 264
Reg. 3(1) 224
(2) 225
Reg. 4 226
Dangerous Substances (Conveyance by Road in Road Tankers and Tank Containers) Regulations (S.I. 1981 No. 1059) 99, 221
Reg. 7 221
Regs. 8, 10 and 21 221
Sched. 1 222

1981 Health and Safety (Dangerous Pathogens) Regulations (1981 S.I. No. 1011) 77
Baking and Sausage Making (Christmas and New Year) Regulations (S.I. 1981 No. 1498) 259
1982 Petroleum-Spirit (Plastic Containers) Regulations (S.I. 1982 No. 630) 259
1982 Hydrogen Cyanide (Fumigation of Buildings) (Amendment) Regulations (S.I. 1982 No. 695) 259
Offices, Shops and Railway Premises Act 1963 etc. (Metrication) Regulations (S.I. 1982 No. 827) 259
Pottery (Health etc.) (Metrication) Regulations (S.I. 1982 No. 877) 259

ABBREVIATIONS

ACAS	Advisory, Conciliation and Arbitration Service.
APAU	Accident Prevention Advisory Unit.
ASTMS	Association of Scientific Technical and Managerial Staffs.
AUEW	Amalgamated Union of Engineering Workers.
BASEEFA	British Approvals Service for Electrical Equipment in Flammable Atmospheres.
CBI	Confederation of British Industry.
EAT	Employment Appeal Tribunal.
EMAS	Employment Medical Advisory Service.
EPEA	Electrical Power Engineers' Association.
FCG's	Field Consultant Groups of the Health and Safety Executive.
HELA	Health and Safety Executive Local Authority Enforcement Liaison Committee.
HMACAI	H.M. Alkali and Clean Air Inspectorate.
HMFI	H.M. Factory Inspectorate.
HSC	Health and Safety Commission.
HSE	Health and Safety Executive.
HSWA	Health and Safety at Work etc. Act 1974.
ILO	International Labour Organisation.
LMRA	Labor Management Relations Act (U.S.A.).
MSC	Manpower Services Commission.
NIOSH	National Institute for Occupational Safety and Health (U.S.A.).
NLRB	National Labor Relations Board (U.S.A.).
OSHA	Occupational Safety and Health Administration (U.S.A.).
PED	Petroleum Engineering Division, Department of Energy.
SOGAT	Society of Graphical and Allied Trades.
SRSCR	Safety Representatives and Safety Committees Regulations 1977.
TLV's	Threshold Limit Values.
TUC	Trades Union Congress.
TULRA	Trade Union and Labour Relations Act 1974.
UKAEA	United Kingdom Atomic Energy Authority.
USDAW	Union of Shop, Distributive and Allied Workers.

CHAPTER 1

INTRODUCTION

The Health and Safety at Work etc. Act 1974, which implemented most of the proposals of the Robens Committee on Safety and Health at Work,[1] represents the culmination of a somewhat tangled story concerning the legal treatment of a central issue in any industrialised society—the health, safety and welfare of its workers. Pound has remarked that the law is better at distributing burdens than in allocating rewards. Because of the nature of legal education today and the day-to-day demands of legal practice, the instinctive reaction of lawyers to the topic "health and safety at work" is to think of remedies (mainly compensation) for injuries which have already been suffered. The common law is concerned with the commutative rights of individuals and with "transfer payments," as the economists put it, and not with the wider social issue in which injuries and disease at work cause huge social and economic losses to society as a whole.

The common law evolved a standard of care measured by reference to the reasonable employer, not in order to prevent accidents (although such might be the result), but to distribute the loss which has occurred, or will occur in the future, as a result of the breach of the duty of reasonable care. It has been said that the standard of reasonable care is a form of subsidy to entrepreneurs in an expanding economy. Indeed, the need to show fault on the part of the employer, together with the further limitations that the employee should not be able to recover for the negligence of a fellow-employee with whom he is in "common employment" (the doctrine of *collaborateur*), nor for risks which he has consented to run (the *volenti* principle), goes to demonstrate that short of "personal" fault on the part of the employer, the worker should really be his own insurer. In a country where gradualism is an article of political faith, attempts were made to rid the common law of some of its asperities towards workers, of which the Employers' Liability Act 1880 was a statutory example. Common employment, consent to risk and the denial of a remedy by reason of the

[1] Report of the Committee 1970–72, Cmnd. 5034.

contributory negligence of the worker were also reformed.[2] It is, however, interesting to perceive the individualistic bias of the common law in a case like *Griffiths* v. *Dudley (Earl of)*[3] in 1882 where it was held that it was not contrary to public policy for a workman to contract out of the benefit of the 1880 Act.

The claim has been made that the common law has a preventive role in addition to its main reparative role. As one judge has remarked, the civil sanctions running into tens of thousands of pounds can be a more coercive sanction than small fines imposed under factory legislation. The dual roles of the common law—reparative and preventive—have incurred some well-deserved criticism. As Professor Atiyah has stated—

> "If we could find a compensation system which served the purposes of compensation well while at the same time it provided a deterrence against conduct which caused accidents or injuries, we would embrace it with enthusiasm. But if we try to achieve these two different objectives—compensation and deterrence—by one regime, we may well end up with worst of all possible worlds."[4]

If the common law has some preventive effect, that effect is variable and easy to exaggerate. Most cases involve negligence rather than deliberate wrongdoing. By definition, negligence is a state of mental inadvertence when the possibility of sanctions is normally furthest from the tortfeasor's mind. The idea that the law of tort is a loss-shifting mechanism between the victim and the tortfeasor, an idea traceable back to Aristotle who saw commutative justice as a method of redressing a balance disturbed by a wrongful act or omission, is largely undermined by the practice of insurance which, in respect of employer's liability, has been compulsory since 1969.[5] Tortious loss is then shared amongst the policyholders. No doubt, premium loading in the light of a bad claims record has some deterrent effect, but even then the economic cost of such loading may be passed on to the consumer as the ultimate guarantor. In its evidence to the Robens Committee, the British Insurance Association accepted that insurers can have considerable influence with regard to loss prevention but warned that there is "obviously a limit to which funds established to provide indemnity against such losses can

[2] Common employment was abolished by s.1(1) Law Reform (Personal Injuries) Act 1948. Apportionment by reason of contributory negligence was substituted by the Law Reform (Contributory Negligence) Act 1945. The *volenti* principle, a doctrine invented by the common law, was circumscribed by later decisions, notably *Smith* v. *Baker & Sons Ltd.* [1891] A.C. 325.

[3] (1882) 9 Q.B.D. 357.

[4] Professor P. S. Atiyah *Accidents, Compensation and the Law,* Weidenfeld and Nicolson, at page 546.

[5] Employers' Liability (Compulsory Insurance) Act 1969.

be used for their reduction, since too great an expenditure in this way would necessarily escalate premiums beyond acceptable limits."[6] Insurers after all are in the indemnity business; not the business of accident prevention. Although we have flirted with the idea of differential social security contributions, the difficulties deterred the Pearson Commission[7] from making an agreed proposal on these lines, so that our social security system continues without any preventive effect inasmuch as the claimant recovers benefit irrespective of "fault" on the part of the "wrongdoer."

One certainty is that the concentration on compensation has distracted attention (including the attention of some trade unions) away from prevention, and the manner in which protective legislation has been invaded by litigants in search of compensation (once the common law had accepted breach of statutory duty as a common law tort) has served to muddy the waters of that legislation. We shall see when we discuss section 47 of the 1974 Act that an attempt has been made to separate the two streams (see below, pp. 172 *et seq.*).

PROTECTIVE LEGISLATION

Early in the nineteenth century it was realised that a law of compensation could not deal with the scale of injury and disease which the new industrialism was bringing in its train. The factory system did not begin with the Industrial Revolution but was rapidly extended by that Revolution. The first of the Factory Acts (42 Geo. 3, c. 73)[8] in 1802 purported to deal with the bad conditions caused by the Poor Laws and Re-Settlement Acts, rather than with the evils of urban industrialism which, as described by Marx and Engels, were yet to come. In a century of *ad hoc* legislation, a series of Factory Acts was passed, starting with textile factories but later extended to other factories, until a measure of rationalisation was introduced by the Factory and Workshop Act 1878, which consolidated and amended factory law. On the Second Reading of that Bill, it was said that—

> "it is desirable, in the interests alike of employers and employed, that all trades and manufacturers employing the same class of labour should be placed upon the same footing, and under the same protective and restrictive regulations,"

to which was added the euphoric opinion that the new

[6] Report, vol. 2, p. 43.
[7] Report, Cmnd. 7054–I, paras. 902–903.
[8] This paternalistic Act, which required the instruction of apprentices in the Christian religion, was not a factory act in the modern sense.

legislation "consolidates and simplifies the law beyond the possibility of misinterpretation." With amendments, this Act furnished the basis of the law until the Factories Act 1961 which still continues in force."

The "code" regulating factories served as a paradigm for later codes regulating mines and quarries (the Mines and Quarries Act 1954), agriculture (Agriculture (Safety, Health and Welfare Provisions) Act 1956), offices shops and railway premises (Offices, Shops and Railway Premises Act 1963), together with other Acts regulating explosives, clean air and nuclear installations. The four main codes were supplemented by a subordinate tier of regulations covering matters of detail, but occasionally used to amend matters in the parent Act. The enactment of the Employment Medical Advisory Service Act 1972 set up a new structure to deal with occupational health, and marks the evolution of the law from a preoccupation with physical injury to a wider concern with the physical and mental health of the worker. The result, as one might expect of a country which had pioneered the Industrial Revolution, was considerable legal regulation of health, safety and welfare at work.

Features of the Codes

The four main codes (together with the particular statutes dealing with specialised matters) exhibit certain features many of which are characteristic of British statute law generally. The dominant aim is the prevention of injury and disease by the use of strict standards, administrative enforcement by a state organ and the imposition of criminal sanctions. In contrast with the generalised and diffuse nature of the common law duty of reasonable care, the statutes are framed so that, in the words of a Minister of Labour, "employers generally will know with certainty what is expected of them,"[9] a point made by Professor Atiyah when he recommends the need for "more detailed guidance as to how to behave."[10] The dilemma of protective legislation is that if it is couched in too general a language, the citizen finds it difficult to regulate his behaviour by reference to the language, whereas if the legislation is too detailed and specific, certainty is achieved at the price of rigidity and loss of personal responsibility on the part of the individual. We shall see, as the Robens Committee itself found out, that the emphasis on certainty led to serious deficiencies in the system.

[9] Commenting on the Factories Acts and judicial decisions thereon when speaking on the Second Reading of the Offices, Shops and Railway Premises Bill 1963.

[10] *Op cit.* p. 550. See the use of "standards" below, pp. 7 *et seq.*

Places, Processes and Persons

The ambit of several of the pre-1974 Acts is limited spatially. Terms such as "factory," "mine," "quarry," "office," and the like are defined, although not exclusively defined, by reference to places. In many of the reported decisions the relevant question which is put is "Where was he or she at the relevant time?" the answer to which question may of itself determine the issue. Thus, it was found necessary in section 116 of the Factories Act 1961 to make special provision for young persons who are employed in the business of a factory "wholly or mainly outside the factory." Events such as those at Flixborough and Seveso in Italy demonstrate that it is no longer possible to say that what goes on in factories or other places of work is of interest only to those who work there. We shall see that one of the main changes in the 1974 Act is the manner in which the "public" dimension of accidents is brought within the law.

Secondly, and largely overlapping with the geographical limitation, is a reliance upon functional criteria, *e.g.* in the case of factories, the need for manual labour in relation to listed processes performed in premises by way of trade or for purposes of gain. Similar functional criteria are used in the other codes and, indeed, in the Agriculture (Safety, Health and Welfare Provisions) Act 1956, the definition of "agriculture" is entirely functional although no one can deny that agriculture can be performed only in space and time. The functionalism of the codes is essentially "physical" in that it deals with the physical process and the safety, health and welfare accompaniments to that process, *e.g.* the guarding of a dangerous part of a machine in a factory. We shall see that since 1974 the law has attempted to deal with the human or ergonomic side of safety, bearing in mind that most accidents are caused by human failures rather than failures of "hardware." One recoils from using phrases such as the "man—machine interface," but people themselves can be the weakest link in the safety chain, as in the nuclear incident at Three Mile Island, Harrisburg, U.S.A.

Thirdly, the statutory duties have traditionally been owed by certain persons to certain persons. Most of the duties in factories' legislation are cast upon the "occupier," a term which is not defined in the legislation, but which for practical purposes will usually (although not always) be the employer. Again, in a general sense it would be true to say that the duties are owed to "employees"; this is not entirely accurate since much depends upon the wording of the relevant enactment (a difficult task when the enactment is silent on the matter). Historically, one of the great achievements of protective laws has been the protection accorded to women and young persons, which had far-reaching consequences for adult male workers inasmuch as employers found it impracticable to operate their enterprises with a dual code of hours and conditions of labour for adult male workers on the one hand and women and young persons on the other hand. This protection continues, although it is a sign of social change that the continuation of the protective laws for women has been critically examined by the Equal Opportunities Commission.[11] We may yet see equality of working conditions between the sexes with due allowances for inescapable differences, *e.g.* the pregnancy of a female.

Literalism

The search for certainty, justified as necessary in statutes which in the final resort are criminally enforceable, has tended to a technical and professional literalism which, if appreciated by the lawyer, is not always relished by the layman who, after all, is preoccupied with running an enterprise as opposed to studies in linquistic philosophy. To take one example. The respondent in *Price* v. *Claudgen*,[12] who was an electrician, received an electric shock whilst repairing a neon lighting installation on a cinema which caused him to fall and suffer an injury. He was held to be outside the pale of the building regulations which covered "repair or maintenance of a building" because the neon sign, whilst "on" the building was not "part" of the building. Perhaps the most famous example is the House of Lords' decision in *Nicholls* v. *Austin (Leyton) Ltd.*[13] which held that the strict duty to fence a dangerous part of a machine under section 14(1) of the Factories Act 1961 did not relate to materials or parts of the machine itself which are ejected from an unfenced machine, the purpose of the section being to prevent contact between the machine and the operator. Several other examples might be cited were it not for the welcome development in *Wearing* v. *Pirelli*[14] in which the House of Lords declined to be drawn further down the path of pedantic literalism. The argument in that case, namely, that because the operator had been injured by contact with a thin rubber sheet on a drum he had not made contact with the drum itself, was sensibly and laudably rejected. The law has to deal with life, not logic, as Chief Justice Holmes has reminded us.

[11] Health and Safety Legislation—*Should we distinguish between men and women*? Equal Opportunities Commission (1979).
[12] [1967] 1 W.L.R. 48.
[13] [1946] A.C. 493 (the common law may afford a remedy where the risk of injury was foreseeable).
[14] [1977] 1 W.L.R. 48.

The sophistry and false logic with which factory legislation has been interpreted is admirably reviewed by Lord Hailsham L.C. in *F.E. Callow (Engineers) Ltd.* v. *Johnson*,[15] where he gives telling examples of protective provisions which, designed for the protection of workmen, afford "protection" which "on closer scrutiny" turns out to be "illusory." His Lordship, however, considered that the die was now cast and that it was too late for the Courts to close the gaps (gaps, it may be noted, which the Courts had opened up); any reform must come from Parliament or from Ministers using their regulatory powers. The problem of literalism is not peculiar to protective laws but affects the increasing flow of statutes seemingly inseparable from parliamentary democracy. The British have set their faces aginst broadly-framed prescriptions in the continental tradition. The British on the whole prefer statutes which are proof against the imbecile who requires detailed guidance and the charlatan bent upon misunderstanding the law. In fairness, the general duties in the Health and Safety at Work Act 1974 have a continental generality which is to be commended. The British judiciary, with one or two noted exceptions, seems to be increasingly worried about its appointed and non-elected status in a democratic society and it is this which may cause judges to seek safe refuge in literalism leaving policy and purposes to the hard-pressed Legislature.

Standards

In contrast to the somewhat permissive common law standard of reasonable care, objectified for our purposes in the standard of care owed by a notional reasonable employer, it was felt that protective legislation would best be served by insistence on absolute standards. The case of *John Summers and Sons Ltd.* v. *Frost*[16] is often cited as the draconian example of the absolute duty. In that case, which turned upon the duty to fence dangerous parts of machinery, a workman had suffered an injury when his thumb came into contact with a power-driven grinding machine. The duty to fence was held to be "unambiguously absolute" and did not allow of the excuse that by leaving a seven-inch aperture to enable the grinding operation to be performed the defendant had done all that he could reasonably have done short of fencing off the whole machine. The decision, acknowledged by those delivering it as being tantamount to

[15] [1971] A.C. 740.

[16] [1955] A.C. 740: but see the Abrasive Wheels Regulations 1970, which provide an escape from the dilemma.

"the destruction of the machine as a working unit," elevates the statutory purpose to protect workers above the economic and physical capacity of employers to comply with legislation. Even when a defence of "impracticability" is available, this may be narrowly construed, as occurred in *Boyton* v. *Willment Bros. Ltd.*[17] in which it was held not to be impracticable to put up a guardrail to protect demolition men who were knocking down a building (one of whom had fallen over the edge of the building), however "impracticable" it might be in the ordinary sense to erect protective guardrails as a building is being knocked down.

Generally speaking, the common law leans against absolute standards which are either impossible or extremely difficult to achieve. *Lex non cogit at impossibilia* (the law does not demand the impossible) represents a maxim applicable to any system claiming to be a legal order. The citizen cannot regulate his conduct according to a law which he cannot read or which lays impossible prescriptions upon him. There is, however, a world of difference between a law requiring citizens to walk on air and a law requiring a manufacturer to make safe products or an employer to avoid overcrowding. Even so, the tendency has been to temper the rigour of absolutism by recourse to concepts such as foreseeability, identification of the type of injury envisaged by an enactment and causation (including hypothetical causation);[18] so that today it is usual to speak of "strict" duties rather than "absolute" duties. It is sometimes said by lovers of the absolute that the judges have "diluted" the rigorous standards intended by the Legislature; but it should be remembered that most of the cases construing the statutes have involved civil claims for compensation where the concept of fault still plays a dominant role (see below, p. 173).

Many statutory duties do not employ standards of strict liability but are in a loose sense nearer to the common law standard of reasonableness. We shall see that the standard used in Part I of the 1974 Act, that is the standard of "reasonable practicability," is less stringent and categorical than the "thou shalt" prescription used in many of the provisions in the codes. As Lord Reid put it—

> "Sometimes the duty imposed is absolute; certain things must be done and it is no defence that it was impossible to prevent an accident because it was caused by a latent defect which could not have been discovered—still less is it a defence to prove that it was impracticable to carry out the

[17] [1971] 1 W.L.R. 1625.

[18] Breach of duty does not cause the injury because the device, etc., required would not have been used had it been provided.

statutory requirement. But in many cases the statutory duty is qualified in one way or another so that no offence is committed if it is impracticable or not reasonably practicable to comply with the duty. Unfortunately there is great variety in the drafting of such provisions. Sometimes the duty is expressed in absolute terms in one section and in another section it is provided that it shall be a defence to prove that it was impracticable or not reasonably practicable to comply with the duty. Sometimes the form adopted is that the occupier shall, so far as reasonably practicable, do certain things. Sometimes it is that the occupier shall take all practicable steps to achieve or prevent a certain result. And there are other provisions which do not exactly fit into any classes. Often it is difficult to find any reason for these differences."[19]

Three main classes of duty were identified during the Committee Stage of the Health and Safety at Work etc. Bill.[20] First, there is so-called absolute duty typified by section 2 of the Factories Act 1961, which lays down a broad duty to avoid overcrowding, which is then particularised by a specific requirement (400 cu.ft. of air-space for each person employed). As Lord McDermott explained in one case, "Once the absolute nature of the duty imposed by the statute is imposed, that is proof enough."[21] Second, there is a duty to achieve a state of affairs or to carry out some precautionary measure unless it is impracticable, such as the duty under section 4 of the Factories Act to render harmless fumes, dust and the like so far as is practicable, by securing and maintaining effective and suitable provision for workrooms. Practicability, which means something more than physically possible, is a more severe test than reasonable practicability. A precaution may be physically possible but may not be practicable in the legal sense. To use American terminology, as used in the Occupational Health and Safety Act 1970, the precaution may be physically possible but not "feasible." Third, there is the test of reasonable practicability, used in pre-1974 statutes (*e.g.* section 29 of the Factories Act 1961) but now given a fundamental role in Part I of the 1974 Act (see discussion of this test at pp. 62 *et seq.*, below).

Whilst some statutory duties, particularly those framed in particular terms (such as the duty to provide and maintain washing facilities in section 58 of the Factories Act 1961), afford little escape to an employer or other person who has not

[19] *Nimmo* v. *Alexander Cowan and Sons Ltd.* [1968] A.C. 107, 113.
[20] Official Report Standing Committee A, Health and Safety at Work, etc., Bill 2 May 1974, Cols. 86 *et seq.*
[21] *Galashiels Gas Co.* v. *O'Donnell* [1949] A.C. 275.

complied with the duty *scriptis literis*, the use of terms such as those used in the 1974 Act—"risk"—"safe"—"harmless"—are unlikely to be interpreted in an absolutist sense. First, such terms are likely to be construed by reference to foreseeability, on the principle that it is unfair and unjust to expect someone to take precautions against an unforeseeable risk, even where some qualifying phrase such as "so far as is reasonably practicable" is not present. Second, the whole purpose of the legislation is to protect against industrial risks so that, for example, an office-worker sitting by an open window who is stung by a bee would have difficulty in convincing a Court or tribunal that his or her working environment was unsafe.

The Role of Law

There is a general acceptance of the need for law to regulate what is a political, social, economic and moral problem—the injuries and fatalities arising from accidents at work. The suffering and losses to the victims alone, without consideration of the losses caused to their families, their employers and society at large, call for the intervention of law in its classic role—the prevention of evil. In its simplest form, the law is directed to human conduct, whether by act or omission. Aquinas, writing in the thirteenth century, defined law as "a rule and measure of acts, whereby man is induced to act or is restrained from acting; for *lex* (law) is derived from *ligare* (to bind) because it binds one to act," the end of law being "ordained to the common good."[22] It may come as a surprise to the layman that the explanation for the binding nature of law, its normative character, is a matter of acute controversy amongst legal thinkers. John Austin in his nineteenth century examination, *The Province of Jurisprudence Determined* (1832), whose philosophical thinking continues the positivist line set out by Thomas Hobbes,[23] has exercised a powerful influence on English juridical thought and probably continues to represent a continuing strand of thinking on law despite the modern criticisms of his idea of law. Austin saw law as a command issuing from a sovereign who can ensure compliance by visiting pain and evil upon those who disobey the command. One is obliged to obey the law for fear of the pain and evil which may befall one if one disobeys. Austin saw the "sanction" as the reason for compliance, although he was at pains not to make his definition of law a predictive statement, namely, that pain and evil would or might necessarily follow a transgression. A few millenialists apart, no legal philosopher

[22] *Summa Theologica,* Question 90.
[23] Hobbes's theory of natural law is difficult to comprehend.

has been able to escape from the need of the State to resort to force to control the lawbreaker. What H.L.A. Hart rightly criticised in his *Concept of Law* was the explanation of a duty stemming from force—the "gunman situation"—when in his view the "oughtness" of a rule, that is its validity, stems from a legal order in which rules have an "internal" and an "external" aspect and which are ultimately validated by a rule of recognition. Why the "insider" with his "internal" view that the rule is binding upon him takes this view is not entirely clear; it might be a cognitive or a volitional point of view, although Hart is clear that the normative validity of a rule does not place the rule before the individual for his personal approval. Several of those who deal with the relationship between law and health and safety at work consciously or subconsciously subscribe to the "imperative" theory of a law commanded under pain of evil, whilst others prefer a "rule-based" approach to law in which legal rules involve attitudinal factors on the part of private citizens and public officials. As Kelsen explained it, a legal norm is in essence a conditional statement consisting of a delict and a sanction; "if p, then q," although Ross considers that law is primarily addressed to public officials, so that, in a sense, all law is necessarily "public law."

The Austinian model, derived from the sardonic aside of Hobbes that "clubs are trumps," is clearly inadequate to work the social transformation required by good health and safety laws and this even if we had unlimited inspectorates and unlimited court facilities to punish the wrongdoers. We now see that the law can operate in ways other than ordaining the conduct of its citizens; it can set up institutions and procedures to regulate health and safety in a dynamic sense.

One of the features of a developed legal order is the complexity of laws (commands, norms or rules), both in their scope and their relationship one with another (Kelsen's "hierarchy of norms"). This is nowhere more true than in health and safety law. At the apex of the legal pyramid we have the general duties in Part I of the 1974 Act, of which section 6(*a*) may be taken as an example. This requires designers, manufacturers, importers and suppliers to ensure so far as is reasonably practicable that an article for use at work is so designed and constructed as to be "safe" and "without risks to health when properly used." The law cannot make life "safe" or banish "risk." Life itself is risky. Even if we give up smoking, go jogging and avoid travelling we might still be hit by a meteorite or irradiated by a super novae. No one in his right mind would build an anti-meteorite shelter or wear a thick lead vest to prevent injury by radiaton. Terms such as "safe" and "risk" are contextual and relativist terms. As Lord

Scarman said in a case concerning the statutory duty of Greater London Council to promote "economic" transport facilities for Greater London, the word "economic" is "a very useful word; chameleon-like, taking its colour from its surroundings."[24] Fortunately, those concerned with health and safety are not left with general words and phrases such as those applicable to London transport, since they have a hierarchy of norms which takes them from the general to the particular; that is from the general duties in the 1974 Act, through Regulations, Approved Codes of Practice, Guidance Notes, Voluntary Standards and a variety of Technical Standards (discussed below at pp. 109 *et seq.*). In a formal sense the descent is a journey from the purely legal to the extra-legal.

The whole system is not simply a static array of norms or rules. As Kelsen reminds us, the law regulates its own creation in a dynamic fashion. In a curious way the system of self-regulation brought in by the 1974 Act makes the addressee of the law a "legislator" in his own right. In the Scottish case of *Armour* v. *Skeen*,[25] a director of roads was convicted of an offence under section 37 of the 1974 Act, which was due to the director's "neglect" by reason of his failure to comply with the authority's "Statement of Safety Policy," the Court rejecting the argument that the "neglect" must relate to a duty imposed by legislation. "In my opinion," said Lord Wheatley, "that is a misconception. Section 37(1) refers to any neglect, and that seems to be to relate to *any* neglect in duty, however constituted, to which the contravention of the safety provisions was attributable." The employer's safety policy required by section 2(3), whilst not "law" in the formal sense, represents a "standard" to which he may be held accountable and to which the public officials may hold him.

The Problem of Efficacy

The enforcing authorities and those concerned with the technical efficiency of the preventive system tend to be more interested in the efficacy of the law than in the validity of the law. Lon L. Fuller expresses a common sentiment when he states that—

> "The law cannot enforce itself. Some human agency must be charged with that responsibility; some official of government has to initiate the proceedings by which the law will be applied. If the discharge of the responsibility is lax,

[24] *Bromley L.B.C.* v. *Greater London Council* [1982] 1 All E.R. 129, 174.
[25] [1977] S.L.T. 71.

> tainted with favouritism, or perverted by corruption, then the law is, in effect modified and re-written in the process of being applied."[26]

Measuring the efficacy of law in terms of social and economic results is notoriously difficult. There is a tendency in a parliamentary democracy to devote tremendous effort to new laws and too little effort to seeing how those laws perform in practice, although the latter aspect is now increasingly attracting attention. After the "plateau" in the statistics which was commented upon by the Robens Committee there does seem to be a discernible trend towards the reduction of accidents and incidence rates per 100,000 at risk in manufacturing industry, *e.g.* fatalities reported to HMFI have declined from 196 in 1975 to 147 in 1979, with a decline in the incidence rate (except for serious (Group 1) accidents).[27] The inferences to be drawn from statistics are likely to be unreliable since they involve the "if hypothesis" or, as H.M. Chief Inspector of Factories acknowledged,—"In general it is too difficult to assess the accidents that have not occurred, and the good health preserved."[28] It is interesting to note that the American counterpart of our Act, their Occupational Health and Safety Act 1970 has been praised for its work in reducing injury and disease whilst simultaneously being castigated for failing to produce the results hoped for by Congress. Changes in technology, such as the use of welding robots (which prevent welding burns), the increasing use of microelectronics, and other technological changes, are difficult to evaluate for their effects on health and safety at work. An economic recession may cause a downturn in the gross accident statistics and (a point not confined to the present subject-matter) changes in the system of reporting accidents may appear to reflect a change which has not in fact occurred. Nevertheless the Chief Inspector of Factories quotes the dramatic drop in power press accidents (from 450 to 48) following the operation of the power press regulations in 1965, saying—

> "The number of power presses in use may of course have declined; there may have been changes in economic activity and some press work may have been done using machines other than orthodox hand fed presses. All these factors together could not however have accounted for this dramatic reduction in accidents."[29]

[26] *Anatomy of Law*, Penguin Books, p. 31.
[27] *Health and Safety, Manufacturing and Service Industries* 1979, HSE Report, H.M.S.O.
[28] *Ibid.* p. vi.
[29] *Ibid.* p. vi.

As experience with race relations shows, one of the most challenging problems presented to law is that of changing attitudes which may be socially entrenched. In the Tyneside shipyards after the Second World War, workers who donned the protective headgear provided by management were occasionally pelted from the upper stagings with lethal nuts and bolts for being "cissy"; fortunately times have changed since then and failure to wear a helmet would now generally be treated as stupid. In the current examination of ways of dealing with major hazards (discussed below pp. 228 *et seq.*), the authors of the Second Report of the Advisory Committee on Major Hazards consider that—

> "Ideally, then, any new regulations should provide that the unwise think again, that the ignorant seek knowledge, and that the incompetent are never in control."[30]

There is a powerful and radical segment of opinion which, seeing a social evil, recommends a law of strict liability as the best legislative panacea. The scale of the social evil is perceived to override the requirement of fault or *culpa* on the part of the person whose conduct needs to be controlled. The canon of strict liability which is advocated by some for product liability and health and safety at work displaces the canon of moral responsibility. There were those who would have liked the 1974 Act to embrace strict liability as opposed to the more permissive standard of reasonable practicability (discussed below at pp. 62 *et seq.*). The Factory Inspectorate (unlike the Mines Inspectorate) has fallen back on the technique of prosecution, and there have been some important prosecutions since 1974 under the Health and Safety at Work Act; but in general, it has preferred advice and persuasion to the big stick of the criminal law. The HSE in its Annual Report for 1979–80 acknowledges that—

> "In raising the level of workplace health and safety awareness, and encouraging a more positive approach to combating hazards, we have proceeded . . . *on a basis of agreement*" (*italics supplied*).

This consensual approach operates at national level using consultation on regulations and codes of practice and relying on the advice of advisory committees), and at plant level (by consulting safety representatives and safety committees). The new summary powers to issue improvement and prohibition notices, backed up by the ultimate discretion to prosecute, are seen as reserve powers to coerce the recalcitrant.

[30] HSC, Second Report, Advisory Committee on Major Hazards, para. 54.

In this respect, the Factories Inspectorate and the Alkali and Clean Air Inspectorate have incurred strong criticism from those who see vigorous prosecution as an index of efficacious law. HMFI were criticised for their complaisant approach to the dangers from lethal concentrations of asbestos at the Bermondsey factory of Central Asbestos which closed in 1969. Seven employees were awarded £86,469 by way of damages.[31] Similarly, at the Acre Mill factory of Cape Asbestos, five hundred thousand pounds was paid in compensation to dependants of one hundred employees who had died of asbestosis.[32] HMACAI incurred censure for their inactivity in relation to lead pollution arising from the RTZ smelter at Avonmouth where many workers were made ill from lead poisoning. Abnormally high concentrations of lead were found in shellfish and in the leaves of trees. There seems to be little doubt that there was *vis inertiae* so far as preventive enforcement was concerned in these much-publicised cases. The Report of the Parliamentary Commissioner for Administration on the enforcement of the Factories Act at Acre Mill, Hebden Bridge, contains serious criticisms of HMFI and the Department of Employment.[33] No one doubts that, where the risks have become known and the issues are serious, vigorous action needs to be taken. What is questionable is the use of legal powers on a large scale across the board as a matter of general policy. It is easy to be wise after the event. The inspectorates are frequently faced with the need to balance enforcement action against the social and economic costs of enforcement. The factory closed down because of some danger represents a loss to employees (notwithstanding guarantee payments), to employers and society; these losses have to be taken into account, a matter which tends to be glossed over by those who take a more absolutist view of safety and health.

It is a postulate of our free enterprise system that a person may engage in business without state *fiat* or permission. Regulatory laws, therefore, affect such a person once he is already in business. An alternative to regulatory law would be some form of licensing system.

Licensing is already used in connection with nuclear energy, drug manufacture, storage of petroleum spirit and explosives, where the risks justify a departure from a regulatory régime. It would be difficult in a free economy to extend the principle across the board to all industrial and commercial activities. One

[31] See *Smith* v. *Central Asbestos Co. Ltd.* [1971] 3 All E.R. 204; *Central Asbestos Co. Ltd.* v. *Dodd* [1972] 2 All E.R. 1135 (H.L.), where Lord Salmon categorised the approach of the factory inspectors as "supine" (p.1156).

[32] Source: *The Listener*, May 2, 1974, pp. 559–562.

[33] See Parl. Deb. H.C., 30 March 1976, cols. 1103–1115.

of the problems confronting the Advisory Committee on Major Hazards is the best method of legal control for major hazards and in its Second Report it rejects a mechanistic system of licensing in favour of what is termed "supervised self-regulation,"[34] under which those responsible for a major hazard would be charged with the task of demonstrating that they have taken the necessary steps to ensure the safety of the operation. The proposed new regulations on notification and hazard surveys have been termed "inductive" because they will require management to work out its own solution, which must then be presented to HSE for approval. Such a system, it is thought, would avoid the rigidity of a simple licensing system where the danger is that the licence once given becomes a franchise for the taking of liberties and risks. Supervised self-regulation lies halfway between licensing proper and detailed regulation. Planning law also provides a form of control, now that HSE is consulted on the health and safety dimensions of planning applications involving hazards.

LAW AND ECONOMIC ANALYSIS

The relationship between certain laws and economic ideas has been accepted for some time. Doctrines such as freedom of contract, or its opposite, restraint of trade, have clearly evolved from a *laissez-faire* theory of economic liberalism. Within the last two decades or so an attempt has been made to subject all law to economic analysis; even the criminal law has been seen as a sort of economic market-place. In one of the leading textbooks on the subject, Professor Richard A. Posner's *Economic Analysis of Law*,[35] the American Occupational Safety and Health Act, which lays down minimum federal standards of worker safety and health, has been subjected to an analysis which suggests that the Act was passed in the interests of neither the workers nor the employers. The employer is interested in providing the optimal (which is not necessarily the highest possible) level of worker health and safety, a level which Posner illustrates by an accident or illness cost of one dollar which can be eliminated for ninety-nine cents. Because the workers presumably demanded compensation for the one dollar expected cost (the market providing a datum point in this connection) the ninety-nine cents expenditure will reduce the wage bill by one cent. Posner sees legislation as raising the level of health and safety beyond the level desired by the employees and the employers "and both

[34] H.M.S.O. (1979), para. 77 on p. 26.
[35] 2nd Edition, 1977, p. 246.

groups will be harmed." Various reasons are proffered for this unusual conclusion—the aim of reducing competition from non-union labour, the correction of inefficiencies resulting from governmental programmes for the support of injured workers, *e.g.* in Britain the costs incurred by the "free" National Health Service in treating injuries and diseases suffered at work.

In the absence of external costs, a perfectly competitive market will be socially efficient, but that market may break down when external costs are imposed on individuals. That the market need not necessarily break down is demonstrated by the so-called Coase Theorem which attempts to show that the existence of harmful effects could be accommodated within a model which sets a socially efficient level for the harmful action, both where the market is perfectly competitive and where the law allocates rights to compensation. The Theorem is only a model which is based upon some artificial assumptions, *e.g.* that the costs in the bargaining process are zero. What Calabresi[36] terms "the theoretical basis of general deterrence" is based upon certain postulates of which the most important "is that no one knows what is best for individuals better than they themselves." So if the cost of activities is not included in the prices of those activities people will choose the more accident-prone activities than would be the case if the prices included accident costs. "But," as Calabresi puts it, "to describe a world of perfect general deterrence is to refute its possibility."[37]

Calabresi took it as axiomatic that the principal function of accident law is to reduce the sum of costs of accidents and the costs of avoiding accidents (leaving out of account what he terms "beneficient side effects" in the provision of a respectable livelihood for lawyers and, one might add, the enforcement "industry" itself). Much of the discussion on accident costs concentrates on secondary accident cost reduction which only comes into play when measures of primary reduction of accident costs have failed. The latter measures, according to Calabresi, may seek either to forbid activities thought to cause accidents or seek to make such activities more expensive. In this taxonomy the British Health and Safety at Work Act is a technique of primary reduction of accidents, *e.g.* the House of Lords decision requiring the fencing of grinding wheels even if that fencing renders the wheels useless for their purpose (see above, p. 7). What Calabresi termed the "second cost reduction" subgoal is not concerned with reducing the number of accidents nor their degree of severity, but is concerned with societal costs, of which

[36] In his historic book, *The Costs of Accidents,* A Legal and Economic Analysis, 1970.
[37] *Ibid.* p. 88.

the main one is compensation of victims after accidents. Most of the emphasis by legal writers has concentrated on the secondary role of accident law, whilst at the same time admittedly, dealing with the preventive effect of a compensatory law.[38]

Doubts have been cast upon the ability of both the old and the new safety laws to provide an incentive for employers to comply with the laws. In the words of one commentator, "the Act depends crucially upon the incentive effect of the *existence* of law rather than [upon] its *enforcement.*" The same commentator draws the inference that, despite the current trends in both theory and practice, "there seems to be evidence that safety improvements have depended primarily on assessment of the economic costs and the benefits involved."[39] The Robens Committee[40] were aware of the economic implications of injury and disease at work and urged that greater knowledge of the cost-effectiveness of preventive measures should be obtained so that scarce resources might be used to the best advantage. The Committee identified prevention costs falling upon the employer, upon the Exchequer and upon the nation as a whole. The first of these—costs falling upon the employer—can be calculated, at least in principle, with little difficulty and would include loss of production, damage to plant, investigation costs, liability insurance premiums and so forth. "If accidents can be made to show up clearly on the balance sheet, the employer can apply the same management effort and technique to accident prevention as he customarily applies to other facets of his business.[41] The costing of accidents and disease, in so far as they affect the Exchequer, yields less easily to precise quantification, as do costs falling upon the nation as a whole. Robens urged the new Authority to undertake research into the economics of accidents and accident prevention.

The difficulties in obtaining reliable information on costs are described in a recent study, which concludes that there are several areas of cost where information is incomplete or totally lacking in any suitable form. But one interesting finding (supported by other research) is the "clear indication that officially recorded injury accidents and cases of prescribed industrial disease alone seriously understate the problem in total," although the study cannot quantify exactly the precise

[38] See Professor P.S. Atiyah *Accidents, Compensation and the Law* and, more recently, *The Economic Approach to Law,* Burrows and Veljanovski.
[39] *Economic Deterrence and the Prevention of Industrial Accidents*, Jenny Phillips (1976) 5 I.L.J. 148–163.
[40] Robens Report (Cmnd. 5034) devotes a Chapter to this subject.
[41] *Ibid.*, para. 419.

amount of understatement.[42] It seems (and this is a point made by HSC/HSE) that a substantial proportion of accidents are minor and unreported. These accidents (which involve slight injury, first-aid or damage only) may account for between one-fifth and three-fifths of the total resource cost of all occupational accidents. The conclusions are that despite recent improvements in health and safety standards and the downward trend in recorded accidents and cases of disease, occupational accidents and diseases in total still represent a significant problem to society and the economy today and that earlier research may have understated the problem.

After some noticeable caution with regard to cost-benefit analysis, the HSC *Plan of Work 1981–82 and Onwards* envisages a greater need to assess the costs and benefits of new statutory requirements now that further improvements are likely to be more costly, whilst, at the same time, confessing that cost-benefit analysis has its limitation. "The comparison between costs and benefits therefore involves value judgments and is not a matter of economic logic; the results of the cost-benefit assessment must be used alongside other criteria in judging the merits of a proposal."[43] All proposed health and safety requirements will as standard practice be subjected to analysis which will be more detailed as the proposals themselves become more detailed.

Risk Appraisal

With the limited resources at their disposal the enforcing authorities have necessarily had to restrict risk appraisal exercises to major hazards (discussed below at pp. 238 *et seq.*). Informal arrangements have existed since 1972 whereby planning authorities are asked to consult HSE before granting planning permission for development on or near major hazard sites (*i.e.* sites in which an accident is likely to cause substantial loss of life or serious injury outside the confines of the workplace). HSE has formed a major hazards risk appraisal group consisting of inspectors with expertise in explosives, fires and chemical hazards.[44] Risk appraisal before accidents occur is obviously preferable to the *ex post facto* inquiries of the Flixborough or Seveso variety. Moreover such an appraisal can express risks in mathematical terms as a matter of mathematical probabilities.

[42] *Costs of Occupational Accidents and Diseases in Great Britain*, Phillip Morgan and Neil Davies, Department of Employment Gazette, November 1981, Vol. 89 No. 11, 477 at p. 483. This valuable article distinguishes between "objective costs" (which ignore transfer payments) and "subjective costs" suffered by the victim and his family.

[43] *Loc. cit.* paras. 113–117, pp. 17–18.

[44] Parl. Deb. H.C. (Written Answers), 15 June 1980, cols. 565–6.

The results may not please everyone, particularly those who live near the increasingly huge storage facilities which we now have. Sir Bernard Braine has consistently questioned the risk at Canvey Island, where successive concentrations of flammable materials constitute in his words, "a national scandal and disgrace."[45] The HSE Report on Canvey Island has been evaluated by an analysis carried out by consulting engineers and scientists, which concludes that the risk calculations are very high and that "the real situation is not necessarily as serious as to expose the Canvey population to excessive risk."[46]

THE MORAL ASPECT

Although we speak of human beings as "ends in themselves" and castigate those who use their fellow-men as "means," the plain fact is that we live in a society which accepts thousands of deaths and injuries every year with resignation if not with equanimity. Some, whose sincerity exceeds their ability, talk of the complete abolition of certain classes of accident, *e.g.* motor accidents; but it is clear that society as a whole does not contemplate total solutions, or even partial solutions which involve unacceptable costs. The "common good" may have unfortunate consequences for the few. Playing rugby football or hang-gliding are permitted despite the risk of injury to those who participate in those activities. We could, of course, as Calabresi has pointed out, prevent accidents by forbidding the activities which give rise to them. We could, in theory, ban sport, circuses, motor-cycling and firework manufacture altogether because they are "anti-social" or even "immoral." Inspectors are empowered to issue prohibition notices to stop the carrying on of activities involving risk of serious personal injury, but this power is restricted to activities to which the relevant statutory provisions apply. The Inspector may prohibit the use of an unguarded machine in a firework factory, but he cannot issue an order addressed to all fireworks manufacturers interdicting the further manufacture of fireworks. There is no room for moral *dirigisme* of this sort in a society claiming to be free. We accept risk as the price of freedom, for the pleasure which it brings or even for the greater risks which are thereby avoided. The circus knife-throwing act is allowed on grounds of pleasure afforded to the spectator. The tunnel built under a busy road involves risks which can be justified by the road accidents

[45] Parl. Deb. H.C., 23rd June 1980, col. 178.
[46] Oyez Intelligence Report; *An Analysis of the Canvey Report*, Cremer and Warner (1980) p. v.

involving pedestrians which will be avoided. This balancing of risk against utility may involve delicate moral problems.

It was formerly fashionable to invoke philosophical utilitarianism to provide some way of arriving at the net balance of convenience or utility either in relation to an individual or, more commonly, a group or nation. The "felicific calculus" is not as popular as was once the case. One of the chief defects of the classical doctrine of utilitarianism (before the doctrine was refined out of existence) was the manner in which it subordinated the few to the aggregate pleasure of the many. The common law seems instinctively to have grasped the amorality of a doctrine which can prefer the happiness or pleasure (not the same things) of the many to the pain of the few, the intensity of whose pain cannot be fed into the calculus. In one famous case,[47] the life of the cabin boy was held to have a value not less valuable than the lives of the two adults who could prolong their existence by eating him. The *maximin* principle of John Rawls (who asserts life as a primary good) is opposed to the sacrifice of the few for the "happiness of the greatest number." Nevertheless, it must be admitted, legal philosophy does not provide us with a definitive answer to all the marginal cases. The manufacture of asbestos brake pads (even with the best precautions) may cause deaths and disease to those who work with that material; but the use of asbestos in brake pads probably saves countless more lives than if an inferior substitute were used. We do not accept this simple "either-or" approach and what we have done is to insist on extremely high and expensive exhaust standards designed to minimise the risk of disease and to bring it within acceptable limits. In other words we compromise.

In a planned society, life is easier to the extent that one can plan out of existence undesired activities. In a free society, planning and compulsion of this sort are unacceptable. Licensing is reserved for installations such as nuclear installations where the safety dimension is overwhelming. By and large persuasion must replace compulsion. Nevertheless the apprehensions of ordinary people have to be taken into account by politicians however unreasonable those apprehensions may be to the technocrat. No longer do we insist that the motor-car must be preceded by the bearer of a red flag. Nowadays, nuclear energy is a cause for public concern, despite an historic safety record which is second to none. Because of the potentially catastrophic effects of a major nuclear incident it is significant that HSC proposes at a time of financial cuts to devote more

[47] *R.* v. *Dudley and Stephens* (1884) 14 Q.B.D. 273.

resources to an increased Nuclear Installations Inspectorate, thereby showing that the public perception of a danger, real or possibly misconceived, is something of which the authorities must take note.

HEALTH AND SAFETY AT WORK AND THE CONTRACT OF EMPLOYMENT

The duties placed upon employers and employees by protective industrial legislation are enforced by administrative sanctions and, ultimately, by criminal sanctions. The duties are cast as duties enforceable in criminal law, although as we have seen, certain of those duties may be enlisted by those who seek civil compensation; in which event it is the law of tort and not the law of contract which is invoked. Greer J. was evidently surprised that anyone should consider the statutory duties in protective legislation as contractual duties[48] and in this he was right, inasmuch as with us health and safety at work is not seen as a matter of contract. Subject to any contrary provisions or the restraints of public policy, there is no objection in principle to a contract for "danger money" or for "dirty money" as recompense for danger or dirty working conditions; but, such express terms apart, protective duties, unlike the employer's duty to take reasonable care for the safety of his employees, have not been pressed into the guise of implied contractual terms. The interaction between health and safety laws and the individual employment relationship has a slightly improvised air in this country, contrasting with the more coherent approach which one finds in the U.S.A. (discussed briefly below)[49] and elsewhere. Nevertheless, it may be useful to touch upon certain aspects of the relationship between occupational health and safety as they bear upon the contract of employment and the employment relationship.

The Common Law

The common law saw the servant as a person subjected to the authority of the master who was entitled to exercise control by ordaining what the employee should do and how he should do it—the What, the How, the When and the Where, as it has been expressed. Even so, it recognised in the servant a limited right to refuse to comply with an order which, in the absence of specific contractual agreement, exposes the servant to a pressing

[48] In *Browning* v. *Crumlin Valley Collieries Ltd.* [1926] 1 K.B. 522, at p. 527.
[49] See p. 29, below.

danger of injury or disease. So, for example, a bank employee who refused to comply with a direction to report temporarily to the bank's head office in Constantinople, was held in *Ottoman Bank* v. *Chakarian*[50] to have justifiably refused to comply with a direction which if complied with would have put him in "real and justified" fear for his life. Nowadays, employment has been emancipated from the shackles of status, but as Kahn-Freund[51] reminds us, it still involves subordination to authority. Health and safety at work is seen as a managerial responsibility with a superadded duty to consult with employees through safety representatives and the like. The servant who exercises his common law right to refuse to comply with an unlawful order is likely to find that his refusal is something of a Pyrrhic victory in that the common law whilst prepared to hold that a refusal is rightful, is not, in the absence of some special circumstance, prepared to order the continuance of the employment either by specific performance or an injunction should his employer decide to sack him for his temerity. In the absence of a lucrative fixed-term contract or a contract for life, the wrongfully dismissed servant must content himself with a limited claim for damages. A "right" to refuse to work is something of a chimera if it is accompanied by the "right" of the employer to dismiss for that refusal.

If the employee is injured he may have a remedy in tort by reason of the employer's breach of duty at common law to provide a safe system of work, safe plant and equipment and proper personnel adequately supervised and, depending upon the severity of the injury, is likely to obtain compensation on a reasonable scale should he choose to sue for it. This duty of care founds an action for damages both in contract and in tort.[52]

The employee for his part must exercise the required skill and due care in the discharge of his duties. The want of either may justify the employer in dismissing him, and may involve the employee in an implied contractual duty to reimburse his employer damages which he has had to pay out to a third party injured by reason of the employee's negligence.

Unfair Dismissal

The vulnerable position of the employee at common law has been considerably improved by the availability since 1971 of a

[50] [1930] A.C. 277 (*cp. Bouzourou* v. *Ottoman Bank* [1930] A.C. 271): *Kaukul* v. *Anglo-Soviet Shipping Co.* (1931) 41 Lloyds L. Rep. 90.
[51] *Labour and the Law*, 2nd. ed., p.3.
[52] *Mathews* v. *Kuwait Bechtel Corporation* [1959] 2 Q.B. 57.

statutory remedy for unfair dismissal.[53] This remedy is essentially compensatory, notwithstanding the statutory pre-eminence given to re-engagement or reinstatement. The latter two are rarely recommended and, when they are, the sanction for breach is an additional and avowedly coercive award of damages over and above the normal compensatory award. The statutory remedy stops short of guaranteeing complete job security to the unfairly dismissed employee. Nevertheless, for most employees the availability and scale of compensation contrasts favourably with the somewhat theoretical rights at common law.

Constructive Dismissal

Health and safety factors may justify the employee in leaving his employment and claiming that he has been dismissed constructively and unfairly. Since the important case of *Western Excavating (ECC) Ltd.* v. *Sharp* in 1978,[54] the test of constructive dismissal has not been the previous test of good industrial relations practice (the "industrial test") but the test of a fundamental contractual breach of contract by the employer (the "contractual test"). However, it now seems clear that the industrial test banished from the front door has crept in at the back door now that the courts and tribunals seem so ready to apply the label of "implied contractual duty" to factors which previously would have been labelled "industrial."[55] An illustration is provided by *Graham Oxley Tool Steels Ltd.* v. *Firth,* in which the employee was held entitled to leave her employment because of the intolerably cold conditions in the shed where she worked although she had made no complaint to her superiors over a considerable period when those conditions obtained. Her employers were adjudged to have broken an implied contractual obligation to provide a proper working environment. So far as the conditions involved a breach of section 3(1) of the Factories Act 1961 (dealing with working temperatures) or of the common law duty of the employer, Talbot J. had this to say—

> "Quite obviously, it would be wrong for any principle to be extracted from this appeal that a breach of the statutory duty, by itself, or a breach of the common law duty, by itself,

[53] See the useful article by Brenda Barrett, *Occupational Safety and the Contract of Employment,* New L.J. October 13, 1977, p. 1011. Not unlike Chakarian (above), the dismissal of an employee (*non obstante* a contract which permitted his posting anywhere in the U.K.) for refusal to work in Belfast was held to be unfair, because "bullets and bombs could, without mercy and without warning, kill, injure or maim at any time."

[54] [1978] Q.B. 761.

[55] Cases such as *Pagano* v. *HGS* [1976] I.R.L.R. 9 (employee rightly refused to drive unsafe van) can now be reincarnated under the contractual test.

would result in a fundamental breach of contract by the employer. In every case, what has to be done is to look at the circumstances of the contract, the obligation under the contract, to determine whether there is a breach of contract and whether the breach is of such a quality that it indicates that the employer no longer intends to be bound by his obligation under that contract."[56]

In an earlier case, the failure of employers to respond to complaints by an employee that she could not wear the goggles provided because she wore spectacles was held to justify the departure of the employee, Phillip J. saying—

"It seems to us that it is plainly the case that employers as part and parcel of that general obligation (to take reasonable care for the safety of their employees), are also under an obligation under the terms of the contract of employment to act reasonably in dealing with matters of safety, which are drawn to their attention by employees, because, unless the matter drawn to their attention or the complaint is obviously not bona fide or is frivolous, it is only by investigating promptly and sensibly individual complaints that they can discharge their general obligation to take reasonable care for the safety of their employees."[57]

This earlier case was distinguished in the Graham Oxley Case where the circumstances were quite different in that the factory in this latter case involved only some fifteen employees and conditions which, apart from any complaint, were known to the managing director. No doubt, the decision in the Graham Oxley Case was right; but it would be a pity if it were to be used as a precedent to enable employees to walk off a job at a time of their own choosing because of some uncommunicated grievance concerning conditions of work, thereby ending their employment, unlike the American "right to refuse" (discussed below) which can only be exercised if the employer has been asked to correct the defective condition.

Fairness

In determining the fairness or otherwise of a dismissal the tribunal is governed by the general test in section 57(3) unless the case falls within one of the "automatic" categories of fairness or unfairness. Health and safety factors—statutory provisions,

[56] [1980] I.R.L.R. 135, at p. 137.
[57] *British Aircraft Corporation Ltd.* v. *Austin* [1978] I.R.L.R. 332, at p. 335.

provisions in Codes of Practice, voluntary standards, common law standards and the like—may be relevant in determining the questions of fairness. Although they may well be relevant, the employee cannot insist that the tribunal or court determine health and safety issues as such. *Lindsay* v. *Dunlop Ltd.*[58] is an instructive case in this respect since it involved the dismissal of the complainant who refused to work in a particular area unless it was cleared of hot rubber fumes. The complainant and other employees had become alarmed by an HSE report on the health dangers to those working in or near hot rubber fumes; but most of the employees and their union had agreed to use masks as a temporary measure pending a further report from HSE. The complainant refused to work with a mask and was dismissed. The working conditions in the factory were, admittedly, "far from ideal," and the concentration of fumes exceeded the threshold limit recommended by the British Rubber Manufacturers' Association; but the management were, albeit belatedly, seeking to guard against a danger to health which was only suspected but not proved in a factory which could only be altered radically at great cost. The EAT considered that it confused and obscured the issue to say that the tribunal ought to make findings on the employer's duty to provide a safe system of work and safe premises or on his duty to deal with dust and fumes as required by section 63(1) of the Factories Act 1961. Breach of the former duty would be the basis for any claim for personal injury whilst breach of the latter duty might lead to criminal prosecution. It was not for the tribunal to determine these issues but to determine the general issue of fairness under section 57(3) of the Employment Protection (Consolidation) Act 1978, although in doing so it might have regard to health and safety requirements.

It may be fair to dismiss an employee who is in breach of health and safety requirements[59] and, as in the case of the employer, legal duties, such as the duty of the employee under section 7 of the 1974 Act, are amongst the factors which may be taken into account. Workers who walk off the job because of a grievance, leaving a high pressure steam system switched on and omitting to inform management when they would return, have been fairly dismissed.[60] Similarly, the dismissal of an employee who refused to have his hair cut, following a warning by a factory inspector to the employer, has been held to be a fair dismissal.[61] In certain occupations, such as that of an airline

[58] [1980] I.R.L.R. 93.
[59] As in *Wilcox* v. *Humphreys & Glasgow Ltd.* [1975] I.C.R. 333 (workman omitted safety check on conversion to North Sea gas).
[60] *Gannon* v. *Firth Ltd.* [1976] I.R.L.R. 415.
[61] *Marsh* v. *Judge International Housewares Ltd.* (1977) COIT No. 511/57.

pilot, train driver, or engineer in a nuclear power station, a single act of carelessness or incompetence might have grave consequences and might render a dismissal fair. The Court of Appeal in *Taylor* v. *Alidair Ltd.*[62] showed itself less forgiving of an airline pilot's error of judgment in landing a plane on a fine day than did an industrial tribunal which had held his dismissal for that error of judgment to be unfair. The Court of Appeal felt that the public safety might override too scrupulous a regard for "due process" and in this it was surely right. Lord Denning M.R. in that case emphasised that the test of fairness is essentially subjective so that if the employer honestly believed on reasonable grounds that the pilot was lacking in flying capability that was a good and sufficient reason to dismiss. It is not for the tribunal to "second guess" the genuinely subjective judgment of the employer if this lies within the "range of reasonable responses" which an employer in that position might have taken.[63]

The Industrial Relations Code of Practice, introduced in 1971, sets out certain considerations to be borne in mind by employers, employees and their representatives with regard to working conditions—

> "Working conditions
>
> 47. The Factories Act and other legislation lay down minimum standards about working conditions. Management should aim at improving on these standards in consultation and co-operation with employees and their representatives.
> 48. Management should therefore take all reasonable steps to:
>   (i) improve standards of 'housekeeping,' including the cleanliness, tidiness, lighting, heating, ventilation and general appearance of the workplace;
>   (ii) reduce noise, strain and monotony as far as practicable;
>   (iii) ensure that hazards are reduced to a minimum and the work done as safely as possible.
> 49. Management and employee representatives should:
>   (i) take all reasonable steps to ensure that employees use protective equipment, for example, guards, safety helmets, goggles and ear defenders, observe the standards laid down by law and co-operate in agreed safety measures;

[62] [1976] I.R.L.R. 420. See *St. Anne's Board Mills Co. Ltd.* v. *Brien* [1973] I.C.R. 444 (where the employer formed the opinion that a particular employee had not been guilty of a safety lapse).

[63] The employer should always investigate and give the employee an opportunity to state a case. If the "offence" is not gross, a warning should have been given.

(ii) make the best use of arrangements for consultation about safety and health.

50. Every employee should:

(i) ensure that he understands the health and safety precautions and observes them;

(ii) make use of protective equipment."

Whilst these provisions in the Code do not confer directly justiciable rights, they will be taken into account by an industrial tribunal which considers them to be relevant to any proceedings before it. Thus, it would be open to an employee who has been dismissed to ask the tribunal to take into account some default by his employer in respect of one or more of paras. 47–49 above, whilst, conversely, the employer could point to para. 50 to defend his dismissal.

Although the employer is required to include in the written statement containing particulars of employment a note relating to any disciplinary rules applicable to the employee, section 1(5) of the Employment Protection (Consolidation) Act 1978 dispenses with the need to include in that statement "rules, disciplinary decisions, grievances or procedures relating to health or safety at work." Presumably, it was felt that the system of safety representatives and safety committees for which the 1974 Act makes provision would look after the health and safety interests of workers (although those interests might not be represented at all in non-unionised firms) and, it may be, that it was felt that section 2(2)(*c*) of the 1974 Act (see below at p. 78) makes adequate provision with regard to information in this respect.

Section 58(1)(b)

By virtue of section 58(1)(*b*) of the Employment Protection (Consolidation) Act 1978, a dismissal will be "automatically" unfair if the reason (or principal reason) for the dismissal was that the employee "had taken, or proposed to take, part at any appropriate time in the activities of an independent trade union." It is quite clear that health and safety at work may properly be one of the activities of an independent trade union, from which it follows that the dismissal of an employee for taking part in such an activity would be automatically unfair and would be so without showing a 26-week qualifying period of service. So, for example, the dismissal of a union official, shop steward or safety representative for participating at an appropriate time in health and safety matters on behalf of an independent trade union would be unfair under this provision.

On the other hand, the individual employee may take it upon himself to complain to his employer about health and safety matters without bringing himself within section 58(1)(*b*) as was the case in *Drew* v. *St. Edmundsbury B.C.*[64] where, in the absence of proof that he was acting in a trade union activity, he was unable to establish a complaint under that provision. Whether or not he would have succeeded on the general issue of fairness had he served for the requisite 26 weeks is uncertain. The tribunal seems to have been satisfied that the employer (a local authority) had fallen behind in their statutory health and safety obligations and it would appear that he had alienated the sympathies of one or two of his superiors who regarded him as a troublemaker. The moral would seem to be that if an employee without the requisite period of service cannot demonstrate that he is taking part in a union activity he ought to utilise existing machinery within the workplace such as the system of safety representatives[65] if such a system exists. If he has the requisite service, there is no reason why the dismissal of an employee for insisting on observance of health and safety measures required by law should not be considered unfair any more than the dismissal of an employee who is asked to operate a fraudulent system of accounting for sales of petrol. The danger is that the employee may be carried away by his desire to become a "one-man safety committee."

The American Approach[66]

A four-fold approach to an employee's right to refuse a hazardous work assignment has been used in the U.S.A. First, section 7 of the National Labor Relations Act guarantees to employees the right to engage in concerted activities for the purpose of collective bargaining or other mutual aid or protection. The Supreme Court has held that employees have a right to strike over health and safety conditions affecting their work.[67] This right is essentially a collective right. The NLRB's conclusion that one employee might refuse to work to vindicate a collective interest has not secured universal approval by courts of appeals. Second, section 502 of the Labor Management Relations Act (LMRA) permits work stoppages related to

[64] [1980] I.R.L.R. 459.

[65] The functions of safety representatives include the investigation of complaints by employees concerning their health, safety and welfare at work (Reg. 4(1)(b), Safety Representatives and Safety Committees Regulations 1977 S.I. No. 500).

[66] *Refusals of Hazardous Work Assignments: A Proposal for a Uniform Standard,* Larry C. Backer, Vol. 81, No. 3, Columbia L.R. April 1981, 544.

[67] *NLRB* v. *Washington Aluminium Co.* 370 U.S. 9 (1962)—employees walked out when heating failed on a bitterly cold day after repeated requests to have the heating system repaired.

"abnormally dangerous conditions of work" as an exception to a no-strike clause in a collective agreement which would otherwise cause the discharge of employees who break the clause to be unprotected under section 7 above. Again, there seems to be some doubt whether or not the section relates only to collective action. Third, with the approval of the Supreme Court,[68] grievance arbitration provisions in collective agreements may be invoked. Finally, the Occupational Safety and Health Act 1970 is considered to confer individual rights upon employees. Section II(*c*) enacts that "No person shall discharge or in any manner discriminate against any employee because such employee has filed any complaint or instituted or caused to be instituted any proceedings . . . or because of the exercise by such employee on behalf of himself or others of any right afforded by this chapter." Acting under this provision, the Secretary for Labour issued a regulation[69] in 1973 which, whilst asserting the general proposition that "there is no right afforded by the Act to walk off the job because of potential unsafe conditions at the workplace" went on to accept that "occasions might arise when an employee is confronted with a choice between not performing assigned tasks or subjecting himself to serious injury or death arising from a hazardous condition at the workplace." If the employee, with no reasonable alternative, refuses in good faith to expose himself to the dangerous condition, he would be protected against subsequent discrimination." The American Supreme Court in *Whirlpool Corporation* v. *Marshall*[70] upheld this regulation as a valid exercise of the regulatory power of the Secretary of Labour. There must be a real danger of death or serious injury and insufficient time to eliminate the danger by normal enforcement procedures. The employee should have asked his employer to correct the dangerous condition and his refusal to work must be in good faith. The employee who feels that he has been discriminated against, *e.g.* by dismissal or suspension has no individual remedy but must file a complaint with Occupational Safety and Health Administration which then investigates the complaint, after which it may file a suit in the district court against the employer. One conspicuous strength of the American remedy (also applicable to arbitration) is the power to order reinstatement of the employee.

A proper subject for collective bargaining?

The definition of a "collective agreement" in section 30(1),

[68] *Gateway Coal Co.* v. *UMW* 414 U.S. 368 (1974).
[69] 29 C.F.R. 1977. 12(b)(1)–(2) (1980).
[70] 445 U.S.I (1980).

which is linked with the definition of a "trade dispute," has as its most important subject, "terms and conditions of employment, or the physical conditions in which any workers are required to work." It might appear therefore that health and safety at work could form the subject-matter of collective bargaining so far as the law is concerned. One important qualification can be quickly disposed of and this is the legal impermissibility of a collective bargain which seeks to reduce or subvert the mandatory standards set in protective legislation. Such a collective bargain would be unenforceable on grounds of public policy in the absence of provisions such as that in the statutory scheme for redundancy which permits a collectively-agreed scheme to replace the statutory scheme subject to certain conditions and ministerial blessing. There is unlikely to be much enlightenment from the cases on this point in view of the liking of trade unionists (and many employers) for a "gentleman's agreement" in preference to an enforceable agreement. Indeed, the legislative standards established by laws, such as the factories acts or the mines regulation acts, are perceived by judges as imperative laws, quite separate from contract or agreement. In *Browning* v. *Crumlin Valley Collieries Ltd.*, it was contended on behalf of miners, who were prevented from going to work by reason of the unsafe condition of the mineshaft, that there was an implied contractual duty on the mineowner to comply with the statutory duties laid upon him by the Coal Mines Act 1911, a contention which seems to have surprised Greer J. who considered that it would not have occurred to anyone "that it was necessary to make the statutory provisions in question part of the contract."[71]

Historically, attitudes towards health and safety at work seem to have undergone an evolution from a stoical acceptance of bad working conditions, to collective bargaining in order to improve those conditions. This activity culminated in statutory intervention, starting with textile mills and ended with the comprehensiveness of the 1974 Act. The Webbs have described the early concentration upon the twin issues of pay and hours of work and the mute acceptance of bad conditions.

> "The wage-earner sells to his employer, not so much muscular energy or mechanical ingenuity, but practically his whole existence during the working day. An overcrowded or badly-ventilated workshop may exhaust his energies; sewer gas or poisonous material may undermine his health; badly-constructed plants or imperfect machinery may maim him or even cut short his day; coarsening

[71] [1921] 1 K.B. 522, at p. 527.

> surroundings may brutalise his life and degrade his character—yet when he accepts employment, he tacitly undertakes to mind whatever machinery, use whatever materials, breathe whatever atmosphere, and endure whatever sights, sounds, and smells he may find in the employer's workshop, however inimical they may be to health and safety."[72]

The same authors found that trade unionists began to demand safe, healthy and comfortable conditions of work from about 1840, although the improvement of such conditions did not begin to be a definite part of trade union policy until 1871 or thereabouts. The main aim of trade unions was found to be the establishment of the "Common Rule" to which all must subscribe, employer and employees alike. This common rule was, in the first instance, to be achieved by what the Webbs termed rather quaintly "The Method of Collective Bargaining." It was soon perceived that collective bargaining could produce uneven results, so that the third stage used "The Method of Legal Enactment," namely legislation to lay down minimum health and safety standards in particular sorts of workplaces and processes. Once legislated, these standards secured observance without the damaging accompaniments of strikes and lock-outs and applied to both strongly and weakly unionised workforces—at least that was the theory. The legislation needed, it was realised, more and more inspectors (a cry still heard today) if it was to be effective. Health and safety were seen to be matters of national concern and, indeed, many of the laws, such as those prohibiting the use of "chimney boys" to clean chimneys, owed much to what the Webbs saw as "middle class sentiment."[73] Legislation however can only provide minimum standards; it may be rigid and legalistic, and it came to be used by workers and their unions not as a weapon of prevention but as a useful aid to civil claims for compensation arising from personal injuries suffered at work. But this neat evolution from stoical acceptance (or Christian resignation) of working conditions, to collective bargaining about such conditions and then, ultimately, to the use of legislation, whilst true in a general sense, conceals the overlap and tension between the role of collective bargaining and legislation in today's industrial society.

A strong body of opinion sees legislation as pre-empting the field of health and safety at work where it applies. Thus, in a

[72] Sydney and Beatrice Webb, *Industrial Democracy* (1898), p. 354.
[73] *Op. cit.* p. 364.

comparative study,[74] the view is expressed that "certain kinds of labour standards, particularly those of a protective nature (*e.g.* measures relating to occupational health and safety, the protection of women and young persons and maternity protection), are considered to be primarily a subject for legislation, rather than for collective bargaining, although they can constitute a vital aspect of the terms and conditions of employment." The Robens Committee considered that—"There is no legitimate scope for 'bargaining' on safety and health issues, but much scope for constructive discussion, joint inspection and participation in working out solutions."[75] The Robens view, now confirmed by the 1974 Act and its sequels, is based upon certain postulates—the community of interests between employers and employees in improving health and safety at work, the primary responsibility of management in this respect and the need for a self-regulating, consultative system in the workplace. The system of safety representatives and safety committees (discussed in detail at pp. 190 *et seq.*) is a system of "consultation" as opposed to "negotiation." Nowhere in the Regulations or Code of Practice is there any mention of "negotiation" on matters of health and safety at work.

These postulates have not escaped criticism. Critics doubt the community of interest between a management concerned with profit and a labour force concerned with safety. The dichotomy of collective bargaining for conflicts of interests (*e.g.* on pay) and joint consultation for matters of common interest (health and safety at work) has been attacked, particularly by extreme advocates of "voluntarism" in labour relations. Subject to the incompetence of the collective agreement to reduce statutory standards, there are those who recommend an industrial relations approach to health and safety at work.[76] However conceptually clear the line may be between "consultation" and "negotiation," the line in practice can be blurred and even non-existent. Whatever the label, what is overtly consultation may in reality be negotiation. Certainly, health and safety issues constantly fall for discussion between employers and workers, or their representatives, who are in a "negotiating situation." Indeed, it is often impossible to segregate substantive issues such as pay from working conditions involving safety. One of the arguments (seemingly falsified by subsequent experience) against the present mineworkers incentive scheme was the alleged risk of increased accidents in mines from incentive schemes to increase productivity—a connection be-

[74] *Collective Bargaining in Industrial Market Economies* ILO Publication (1973), p. 143.
[75] Report, Cmnd. 5034, para. 66 (p.21).
[76] David Lewis *An Industrial Relations Approach*, Vol. 3, no. 2, I.L.J. (June 1973), p. 96.

tween pay and safety not unknown in the motor industry where the "speed of the line" can be a controversial issue.

REFORM

Although the long-term trends up to 1970 showed a commendable decline in the number of fatal accidents at work, the decade prior to 1970 showed an increase in the number of accidents both fatal and non-fatal at work, a state of affairs which caused considerable disquiet notwithstanding improvements in the incidence and frequency rates of accidents per 1000 people employed. The emergence of new toxic substances and the increased risk of catastrophic accidents resulting from the scale of industry or the storage of vast amounts of fuels and the like were felt to justify a new overall look at health and safety laws which had been built up piece-meal over the years in the best (or worst) traditions of British gradualism.

These doubts led to the appointment in 1970 of the Committee on Safety and Health at Work under the Chairmanship of Lord Robens. The Committee reported in July 1972[77] with severe but justified criticisms of the old piece-meal system. Two of the chief defects which were found in the system were as follows. First was the problem of apathy at work. The system tended to encourage people to think of health and safety as primarily a matter of detailed regulation by external agencies. There were those who, like the speeding motorist, thought that the only real crime was the crime of being found out.

Second, the statutory system was unsatisfactory on a number of counts. It was felt that too much law already existed in the form of statutes and subordinate legislation all of which was beginning to have a counter-productive effect. More law and more inspectors was not considered to be the answer. The piece-meal nature of the system, admittedly often reflecting the exigencies of particular industries, grew up to deal with particular empirical problems, often physical problems, such as the safeguarding of machinery. This haphazard approach tended to neglect the equally important (and probably more important) questions of attitudes, capacities and performance, or the organisational system within which health and safety is set. The law was felt to be intricate and difficult to understand. It is interesting that what was seen as the merit of the statutory system, namely, allowing employers generally to know with certainty what was expected of them, became a defect of the system. The result was complexity and obscurity. A layman

[77] Cmnd. 5034.

reading some of the judgments of the courts, such as those on the meaning of "in motion" or "in use" applied to a machine, could be forgiven for thinking that he was reading an essay in linguistic philosophy. Above all, the statutory system was failing to keep pace with the new range of problems which accompanied what has been termed the "second industrial revolution."

Finally, the Committee found that the administration of the old system was excessively fragmented, involving as it did different inspectorates under different Ministries.

The diagnosis of the Robens Committee has not escaped some criticism. Some[78] consider that it is not apathy which is to blame but the cost of health and safety measures which is the real problem; whilst others[79] are critical of the assumptions of the Committee, *e.g.* that coercion, compulsion and regulations are largely irrelevant and that apathy can be dispelled by improved knowledge and technical skill. The efficacy of self-regulation has been doubted compared to a policy of rigorous enforcement of the criminal law by a strengthened inspectorate.

The Committee recommended that reform should be aimed at two fundamental and closely related objectives, *viz.* (1) improvement of the statutory system to increase the efficiency of the State's contribution to safety and health at work, and (2) a framework for a more self-regulating system at the place of work. Indeed, the Committee considered that: "The primary responsibility for doing something about the present levels of occupational accidents and disease lies with those who create the risks and those who work with them."[80] This "self-regulation" would operate within a better framework of law.

Against the background of these objectives, the Committee recommended fundamental reforms which provide the basis for the HSWA 1974, although the Act goes further in several respects than the Robens recommendations. These recommendations will be referred to at several places in this book in order to illuminate the provisions of the 1974 Act but it might be useful to set out here the broad drift of the Robens Report.

First, the existing statutory provisions should be replaced by a comprehensive and orderly set of revised provisions under a new enabling Act. This new Act should contain a clear statement of the basic principles of safety responsibility and should be supported by regulation and non-statutory codes of practice (preferably the latter). We shall see that the new enabling Act

[78] Jenny Phillips, *Economic Deterrence and the Prevention of Industrial Accidents,* I.L.J. Vol. 5 (September 1976), 148, 153.
[79] Anthony D. Woolf, Robens Report—*The Wrong Approach*? I.L.J. vol. 2, (June 1973), pp. 88–95.
[80] Report, para. 28 *op. cit.*

contains the "basic principles" in the form of "general duties," notwithstanding the written evidence of the TUC which considered that a general duty cannot provide adequate protection against a wide range of serious risks. Needless to say it has proved impossible to replace at the stroke of a pen the thousands of statutory provisions in the old Codes and subordinate legislation with new and simpler regulations, although it is hoped that in time this Herculean task will be accomplished, a hope now receding at a time of public expenditure cuts.

Second, the narrow clientèles of the old laws should be broadened to include all employers and employees, subject to specific exclusions, and to those self-employed persons whose acts or omissions endanger other workers (employed or self-employed). A novel feature is the recommendation that legislation should take account of the interests of the public in safety and health matters.

Third, unified administration under a new Authority should replace the separate inspectorates under different Ministries. Introducing the HSW Bill, the Secretary of State for Employment, Mr. Michael Foot, referred to "a prolonged and intensive period of inter-departmental consultation" by which he meant that there had been "a first-class Whitehall row"—a reference to rivalry between Ministries reluctant to surrender functions to the Department of Employment.[81]

Fourth, whilst acknowledging in their own words that "any idea that standards generally should be rigorously enforced through the extensive use of legal sanctions is one that runs counter to our general philosophy," the Committee did not abandon criminal enforcement, but felt that such enforcement should be reserved for serious infringements which should then be attended by "exemplary punishment."

Fifth, in those cases where advice and persuasion are unavailing inspectors should have powers to issue improvement and prohibition notices against which appeal would lie to the industrial tribunals.

Lord Robens estimated that a 50 per cent. reduction in accidents would be achieved in five years if the guidelines and recommendations of the Committee were followed.[82] It is difficult to correlate accident figures entirely to legal and institutional changes although, leaving on one side the accuracy or otherwise of the 50 per cent. estimate, it is undeniable that the overall decline in both fatal and total reported accidents over

[81] See below p. 48 for the outcome of the row.
[82] Professor Sir John Wood, Vice-Chairman of the Robens Committee, on *The Robens Committee on Health and Safety at Work,* Conference Papers, Institution of Safety Officers, p. 8.

the Seventies has continued into the Eighties.[83] What does seem to be doubtful is the realisation of the Robens hope for more self-regulation with regulations as a last resort, inasmuch as we are seeing a plethora of regulations which in time will produce a complex régime of law—more law, not less. On the balance side, HSC suggests from the fact that the widespread appointment of safety officers has not been accompanied by any significant rise in the number of complaints to HSE inspectorates that mutually agreed solutions are generally being found to health and safety problems through joint discussion at the workplace.[84]

APPLICATION OF THE ACT

(a) *Territorial*

The 1974 Act applies to England and Wales, and, with the exception of Part III (less section 75) and Schedule 5, to Scotland. It does not apply to Northern Ireland, save to permit regulations to be made which apply in the Province. Under this power, the Health and Safety at Work (Northern Ireland) Order (S.I. 1978, No. 1039 (N.I.9)) has been made in terms modelled on the Act, with certain modifications, *e.g.* enforcement through a Health and Safety Agency for Northern Ireland.

The Act itself is not expressed to have extra-territorial effect, although section 84(3)–(6) permits Parts I and II to be extended by Order in Council, with accompanying court jurisdiction in criminal and civil cases in respect of the extended coverage. Disquiet concerning health and safety on oil and gas rigs, particularly in relation to diving operations and pipeline laying, led to the making of the Health and Safety at Work, etc., Act 1974 (Application outside Great Britain) Order, (S.I. 1977 No. 1232), which came into operation on September 1, 1977. These regulations extend the statutory protection to offshore installations and pipelines within territorial waters or areas designated under the Continental Shelf Act 1964.

The administrative responsibility for offshore safety has been a source of continuing and bitter controversy, with three governmental bodies involved, *viz.* the Health and Safety Commission, the Department of Energy (whose petroleum engineering inspectorate have expertise in problems peculiar to oil rigs, *e.g.* "blow-outs") and the Department of Trade. The Burgoyne Report on Offshore Safety[85] (transgressing somewhat

[83] HSC Report 1980–81, para. 2 and Appendix 9.
[84] *Ibid.* para. 10.
[85] (1979) Cmnd. 7866.

its terms of reference) recommended a single agency and, by a majority, recommended the Petroleum Engineering Division of the Department of Energy as the enforcing authority responsible for structural safety, technical aspects and occupational health and safety under the 1974 Act (when PED would act under an agency agreement with HSE).

The Government, notwithstanding the preference of the minority of the Committee for enforcement by HSE, have accepted the main proposal that the PED of the Department of Energy will be responsible *de facto* but will report to the Secretary of State through the HSE and HSC. Most of the Burgoyne proposals will have been implemented by the end of 1982.[86]

Doubts have been expressed as to the desirability of the Government's plan. In the first place, many feel that there is a conflict between the Department of Energy's roles as oil-sponsoring department (with the objective of bringing oil onstream) and as a safety department. Nevertheless, the expertise of the petroleum engineering inspectorate could have been retained even though transferred to HSE. The argument concerning expertise was fought and lost with the mines and quarries inspectorate (formerly under the Department of Energy), the alkali inspectorate (formerly under the Department of Environment) and the agriculture inspectorate (formerly under the Ministry of Agriculture). Some demarcation problems will still remain, *e.g.* onshore construction of rigs and pipelines will remain an HSE function, whilst ships and submersibles will be under the tutelage of the Department of Trade. The Government considers that PED would retain responsibility in connection with the Mineral Workings (Offshore Installations) Act 1971 and the Petroleum Submarine Pipe-Lines Act 1975.

The particular concern for the safety of divers led to the making of the Diving Operations at Work Regulations (S.I. 1981, No. 399) which require divers to have certificates of training and of medical fitness to dive. Diving operations must take place under a diving contractor who must appoint one or more diving supervisors. The regulations consolidate the law from several sources and make use of the concept of "reasonable practicability" in several of the pure safety duties, *e.g.* that the diving contractor should ensure compliance with the Regulations "as far as is reasonably practicable."

(b) *Personal*

The 1974 Act ushered in a twofold extension of protective coverage, first by including within its ambit virtually the whole

[86] See Parl. Deb. H.C., November 6, 1980, cols. 1472–1546, and December 18, 1981, cols. 255–56.

working population of some 25 million persons (domestic servants and their employers in private households excepted), and, second, by including the public with respect to hazards arising from the activities of persons at work. Because the "new entrants" (estimated at some 8 million persons) were not previously covered by protective legislation, the 1974 Act will provide their sole shield apart from new regulations made under the Act. When we come to the "general duties" we shall see that the Act breaks free from the geographical and functional limitations which have hitherto confined protective laws within defined boundaries.

The Act and regulations made thereunder bind the Crown by virtue of section 48, the Crown for this purpose denoting the departments of state and other bodies acting on behalf of the Crown, but excluding the Crown in the personal sense of that term. The section takes account of the peculiar constitutional position of civil servants by treating those who are in the service of the Crown as though they were "employees of the Crown." Admiration for this example of the Rule of Law must, however, be tempered by the qualifications placed upon Crown liability. Few would argue against the criminal immunity of the Crown from the criminal liabilities regulated by sections 33 to 42, although this immunity does not extend to individual "employees" of the Crown who are guilty of offences as individuals (see p. 169 below). The HSE inspectors have been instructed that Crown servants should not be prosecuted in lieu of the Crown, whilst guidance has also been issued to local authorities on the enforcement of the Act against the Executive itself and its employees; the latter should not be prosecuted in lieu of the Executive itself.

More questionable is the dispensation of the Crown from the enforcement machinery consisting of improvement and prohibition notices (with their appeal arrangements) and the seizure powers, all contained in ss.21–25 (see pp. 136 *et seq*., below), a dispensation which carries with it immunity from the criminal "back-up" sanctions behind the notices. As an interim measure in June 1978, a system of Crown Enforcement Notices was introduced. HSE inspectors may serve these on Crown bodies in circumstances where a statutory improvement or prohibition notice would have been served on an employer in the private sector. The same procedure is used, except that there is no formal appeals procedure. Any failure to comply by the employer leads to approaches by the Executive to higher authorities in the Crown bodies concerned.

CHAPTER 2

THE NEW INSTITUTIONAL STRUCTURE

THE COMMISSION

The Robens Committee recommended a new National Authority for Safety and Health at work which would have "comprehensive responsibility for the promotion of safety and health at work" and which would have responsibility for—

(*a*) the provision of advice to all concerned with safety and health at work;
(*b*) the management of statutory inspection and advisory services, including scientific and technical research facilities and institutions;
(*c*) the setting and reviewing of standards;
(*d*) the acquisition and provision of information and the promotion of research, education and training; and
(*e*) collaboration with national and international bodies concerned with safety and health.[1]

The new Authority would be directed by an executive Managing Board with a full-time Chairman with an authoritative voice in safety and health matters. The members of the Authority would be non-executive, part-time, paid members, who would reflect the interests of both sides of industry, local authorities, and others concerned with safety and health at work. The Authority would be advised on particular subjects by a number of expert advisory bodies.

Moving the Second Reading of the Health and Safety at Work Bill, the Secretary of State for Employment argued that the setting up of the Commission as a centralised co-ordinating body was the most important part of the Bill because the bodies previously responsible for the enforcement of protective legislation in this country were too fragmented, in that—

> "There were five departments involved, nine statutes, five hundred subordinate instruments and seven different in-

[1] Committee on Safety and Health at Work, Cmnd. 5034–(1972), p. 36, paras. 115–116.

> spectorates—all trying to deal with this complicated modern problem that is changing all the time."[2]

The Robens Committee had already recommended unified administration for several reasons, one of which was the need for wide consultation between the different bodies, a situation in which, possibly, all move forward at the pace of the slowest.[3]

The Secretary of State assured the House that the Commission would be responsible through Ministers to the House of Commons, that regulations would be the responsibility of Ministers (particularly in the Department of Employment) and that—

> "it was not the Government's intention to set up a body which was remote from questioning in the House of Commons."[4]

Mr. William Whitelaw, replying to the debate for the Opposition, was convinced that it was right to set up a unified Commission, although he had a doubt concerning the loss of direct ministerial responsibility which hiving off necessarily entails.[5] Mr. Cyril Smith for the Liberal Party also welcomed the formation of the Commission, agreeing that it was a good thing to see nearly all the inspectors combined under one roof, but regretting that the Government was not prepared to include the National Radiological Protection Board and, with regard to the Nuclear Installations Act, the United Kingdom Atomic Energy Authority.[6] So far as the former Board is concerned, the Government felt that it was inappropriate to include it because it has a number of responsibilities outside the scope of health and safety at work whilst, in so far as it has health and safety responsibilities, there is a provision (now section 77) imposing a duty to consult and liaise between the National Radiological Protection Board and the Commission. The Government also pointed out that the Nuclear Installations Inspectorate is included within the oversight of the Commission.[7]

Section 10 provides for the establishment of the Commission, which is to be a body corporate consisting of a Chairman and not less than six nor more than nine members, all appointed by the Secretary of State, the members being appointed after consultations with organisations representing employers, employees,

[2] Rt. Hon. Michael Foot M.P., H.C. Deb., April 3, 1974, Vol. 871, col. 1290.
[3] Report, pp. 11–12.
[4] H.C. Deb. April 3, 1974, Vol. 871, col. 1291.
[5] *Ibid.* col. 1305.
[6] Parl. Deb. H.C., April 3, 1974, col. 1321.
[7] Parl. Deb. H.C., April 3, 1974, col. 1390.

local authorities and other organisations including professional bodies. The first (and present) Chairman was Mr. William Simpson, formerly General Secretary of the Foundry Section of the Amalgamated Union of Engineering Workers and sometime Chairman of the Labour Party. Criticised by some for weakness, he has chosen to proceed by consensus, thereby earning the trust of the two major political parties each of whom has re-appointed him as Chairman during their respective periods in government. The Commission is charged with four general functions under section 11(2), *viz.*—

(*a*) to assist and encourage persons concerned with the general purposes of Part I of the Act to further those purposes;
(*b*) to make such arrangements as it considers appropriate for the carrying out, publishing and encouraging of research and the provision of training and information in connection with those purposes and to encourage research and the provision of training and information in that connection by others[8];
(*c*) to make such arrangements as it considers appropriate for securing that government departments, employers, employees, organisations representing employers and employees respectively, and other persons concerned with matters relevant to any of those purposes and provided with an information and advisory service and are kept informed of and adequately advised on such matters;
(*d*) to prepare regulations.

Although the functions of the Commission (and those of the Executive; see below) and its officers and servants are, by virtue of section 11(7), "performed on behalf of the Crown," the Commission is, in the current argot, a QUANGO, or quasi-autonomous non-governmental organisation, entrusted with a large degree of autonomy in the performance of its diverse functions. The whole purpose of setting up a QUANGO is to devolve functions to an independent body which can bring the necessary specialisation and expertise to bear. Even so, as indicated above, there are fears that this devolution may lead to some loss of ministerial responsibility. To the extent that the Commission is subject to control by a Minister it is "quasi-autonomous." It must submit particulars of what it proposes to do to the Secretary of State, to whom must be given a statement of accounts and a report on the performance of its functions

[8] See John H. Locke on provision of information and advice for the protection of health and safety at work, ASLIB Proceedings, 1976, pp. 9–16, and H.J. Dunster, *So Far As Is Reasonably Practicable* (1976).

during each accounting year.[9] The Minister is something more than a "rubber-stamp" in relation to proposed regulations in that he may modify these after consulting the Commission.[10] He may interdict the approval of a Code of Practice by withholding his consent under section 16(2) or forbid under section 14(2)(*b*) the holding of an Inquiry without his consent.

In the last resort, he has the powerful weapon put into his hands by section 12(*b*), namely, to give the Commission general directions with respect to its functions (including directions modifying its functions, but not directions conferring on it functions, other than those it was deprived of by previous directions given by virtue of the present provision) and any other directions which it appears to him requisite or expedient to give in the interests of the safety of the State. This is a common-form power conferred on Ministers in several acts, including the nationalisation statutes, which is little used and, in the case of the Commission, it would seem, never used.[11] The power to give general directions has been little used in the nationalised industries for the simple reason that Ministers have other ways of securing compliance with their wishes (often political wishes), not least of which is the power of appointment or non-renewal of appointment. It has been said that the "luncheon-table directive" is commoner than the formal direction, possibly because the Minister is more openly answerable to Parliament for the latter than for the former. The Commission normally reports to the Secretary of State for Employment, but is answerable to the Secretary of State for the Environment in relation to pollution.[12]

It would seem that the Commission is more powerful and more effectively organised than its American equivalent the Occupational Safety and Health Administration (OSHA) which is under the aegis of the U.S. Department of Labour. The Secretary of Labour promulgates standards on the advice of the Occupational Safety and Health Administration, whereas the National Institute for Occupational Safety and Health (NIOSH), under the Department of Health, Education and Welfare, is responsible for research education and training. Although the work of OSHA has been praised on this side of the Atlantic, particularly for research and the enunciation of standards for carcinogens and asbestos, American critics are less pleased with NIOSH whose record according to one critic "invites a sense of despair." The same critic notes the slowness of OSHA in promulgating

[9] s. 11(3)(*a*) and Schedule 2.
[10] s. 50.
[11] At least up to 1976, Parl. Deb. H.C., August 6, 1976, col. 1131.
[12] HSC Report, 1974–76, p.3.

new standards,[13] a matter on which the Occupational Safety and Health Act Review (1974) has commented, noting that only three standards had been promulgated on hazardous substances in three years whilst documents submitted some two years previously remained unstudied.[14] It seems that OSHA may face a more daunting task than its British counterpart in that it has to deal with the Federal-State dichotomy and a climate of opinion not entirely sympathetic to interventionist and "do-good" governmental regulation.[15]

The British Commission has a number of detailed powers which are listed in section 13. It may make arrangements and agreements with government departments or other persons for the performance of Commission functions either with or without payment and, conversely, (with one exception) it may make agreements with Ministers, government departments or other public authorities to perform on their behalf functions which in the opinion of the Secretary of State can appropriately be performed by the Commission. The exception relates to a ministerial or departmental power to make regulations or other instruments of a legislative character. Section 77 of the Act enables the National Radiological Protection Board to carry out work on behalf of the Commission. The Commission has entered into a number of agency agreements: with the Secretary of State for Scotland so that the Scottish Industrial Pollution Inspectorate can enforce regulations relating to the control of emissions into the air[16]; and, also, agreements which provide for the enforcement of offshore safety by the Petroleum Engineering Division and the Pipelines Inspectorate of the Department of Energy. Another agreement applicable to premises occupied by British Rail (except railway workshops) provides for the enforcement of legislation in those premises by the Railways Inspectorate of the Department of Transport.

Section 13(1)(*d*) allows for the appointment of persons or committees of persons to provide the Commission with advice in connection with its functions. For this purpose and the remaining purposes set out in section 13, the Commission may make payments at approved rates. This provision implements the Robens suggestion that technical working parties should be set up with the necessary expertise to deal with the detail of

[13] Joseph A. Page and Peter N. Munsing, Law and Contemporary Problems, 1973–74, Vol. 38, at p.654 and p.656.

[14] Hearings before the Sub-Committee on Labor of the Committee on Labor and Public Welfare, United States Senate, November 1974: 93rd. Congress, Second Session.

[15] See judicial review of standards, below p. 121.

[16] See New Law Journal, January 16, 1975, Vol. 125, p.62.

individual regulations, codes and standards.[17] To date, a complex structure of advisory committees has been set up using senior persons from both sides of industry, local authorities and educational bodies, with remits to consider problems related to particular hazards or industries. The current structure embraces seventeen Advisory Committees, namely, three subject advisory committees, three committees concerned with particular hazards, nine industry advisory committees, the British Approvals Service for Electrical Equipment in Flammable Atmospheres (BASEEFA), the Advisory Council and the Safety in Mines Research Advisory Board. Questioned as to the value of its advisory committees by Sir Leo Pliatzky in his examination of non-departmental bodies, the Commission has replied that it attaches great importance to the work of the advisory committees as "one of the most important, and most cost effective, ways of putting into practice the underlying philosophy behind the HSW Act—the involvement of all concerned in the reduction of occupational hazards."[18] Some advisory committees which complete their work may be wound up (as has happened with the Advisory Committee on Asbestos) whilst other committees will no doubt be established in the future.

Section 14[19] gives to the Commission power to direct investigations and inquiries into any accident, occurrence, situation or other matter whatsoever which it thinks is necessary or expedient to investigate for any of the general purposes in Part I of the Act or with a view to making regulations for those purposes. It is immaterial whether the Executive is or is not responsible for securing the enforcement of such (if any) of the relevant statutory provisions as relate to the matter in question.[20] The Commission may direct the Executive or any other person to investigate the matter[21] and make a special report or, with the consent of the Secretary of State, may direct an inquiry to be held into any such matter. The latter (the inquiry) is subject to the Health and Safety Inquiries (Procedure) Regulations 1975 (1975 S.I. No. 335), which have been

[17] Report (Cmnd. 5034), p.49. The decision was made at an early stage that the Commission would not have a large administrative staff but would rely on the Executive to provide the policy staff required: V.G. Munns, *Function of the Health and Safety Commission,* Royal Society of Health Lecture.

[18] HSC Report, 1979–80, p.8.

[19] See Bill No. 241. H.C. Session 1974–75 (an unsuccessful private member's bill to amend this section).

[20] See, for example, Report on Explosion at Houghton Main Colliery, Yorkshire, June 1975, HSE, HMSO 1976.

[21] See Investigation into risks of existing and proposed installations on Canvey Island (*The Times,* March 24, 1976) and The Explosion at Laporte Industries Ltd., Ilford, April 5, 1975, HSE, HMSO, 1976 (this latter accident was investigated by H.M. Superintending Inspector of Factories, London and Home Counties (East) Division).

made under section 14(3), and which require that the proceedings shall be in public, except where the regulations otherwise provide. In pursuance of the express powers conferred by section 14(4), these Regulations specify powers of entry and inspection, the summoning of witnesses and the production of documents, notification of the inquiry (including appearance and representation at the inquiry), and procedure at the inquiry (including a power to take evidence on oath).

The Commission may cause a report of an investigation or inquiry, or so much of it as the Commission thinks fit, to be made public at such time and in such manner as the Commission thinks fit. This provision (now section 14(5) and (6)) attracted criticism at the Bill's Committee stage.[22]

The two-tier constitution of a Commission and an Executive echoes the arrangements in the Transport Act 1968 for a Passenger Transport Authority and a Passenger Transport Executive. An appointed Commission served by an executive arm has certain advantages; the "lay" element puts a brake on rule by technocrats and makes for some degree of democratic control. The cost of a divided system need not necessarily be high. The HSC Report for 1980–81 shows that the Commission had eight members[23] in addition to the Chairman, so that with a small administrative staff its expenditure was only £127,000 in the year ending March 31, 1981. This is tiny compared to the expenditure by the HSE, although it should be added that the budget for 1982–83 is reduced by 6½ per cent. from the 1979–80 level in respect of staff-related expenditure.

THE EXECUTIVE

The Health and Safety at Work etc. Bill received the Royal Assent on July 31, 1974 and those parts of the Act relating to the establishment and functions of the Health and Safety Executive and to enforcement were brought into effect by a commencement order.[24] The Robens Committee[25] had argued strongly for the creation of a single and unified safety and health inspectorate which would bring about improved co-ordination, operational efficiency, and the necessary assimilation of scientific and technical expertise, together with support facilities. It was with this end in view that the Committee therefore recommended

[22] Standing Committee A, Health and Safety at Work etc. Bill, Thursday, May 9, 1974, col. 169, Mr. Robert Cryer M.P. (a dogged exponent of strict liability and systematic inspection).
[23] There is provision for a ninth member
[24] Health and Safety at Work etc. Order 1974 (S.I. 1974 No. 1439 (C.26)).
[25] Report, Cmnd. 5034, p.63.

that the existing inspectorates for factories, mines, agriculture, explosives, nuclear installations, and alkali works should be merged and operate in future under the control of the Commission. The Committee had some reservations concerning agriculture in which problems of organisation occur because "farming units are small, numerous and very widely scattered," for which reason it recommended that, whilst all the full-time agricultural personnel should be brought within the unified inspectorate, the field officers of the Agriculture Department should remain within the Ministry but operate as agents of the new Authority.[26] In fact, it may be interpolated here, that the 1974 Act originally made all matters relating exclusively to agriculture the responsibility of the Agriculture Ministers; but the Employment Protection Act 1975, s.116, provided that those responsibilities should be transferred to the Commission.

Considerable importance was attached to the organisation of a separate multidisciplinary research and development division which would include the Safety in Mines Research Establishment, the Factory Inspectorates Accident Prevention Studies Unit and a Statistics and Economics Department. The emphasis behind the whole re-organisation envisaged by the Committee was on a cost-effective and imaginative use of manpower and scarce resources.

In the event, the recommendation of the Committee was not endorsed by Parliament, the Secretary of State for Employment[27] making it clear on Second Reading that the separate identity of each inspectorate would be retained, a provision for which he was to receive all-party support.[28] The Government was put under considerable pressure to maintain the separate identities of the existing inspectorates, particularly from the National Union of Mineworkers. The Committee[29] had recommended that the new inspectorate should have as its prime objective the prevention of accidents and ill health, and the promotion of progressively better standards at work through the provision of information and skilled advice to industry and commerce. The work of inspection should continue to include as inseparable elements the provision of advice as well as the enforcement of sanctions.

To meet the new demands made on them, the new inspectors would need to be specially trained to enable them to discharge a wide range of duties. There should be a variety of grades and a

[26] *Ibid.*, p.68.
[27] Rt. Hon. Michael Foot, M.P.
[28] Rt. Hon. William Whitelaw M.P. for H.M. Opposition (Parl. Deb. H.C. April 3, 1974, col. 1304); and Mr. Cyril Smith M.P. for the Liberal Party (*ibid.* col. 1320).
[29] Report, p.66.

high degree of in-service training for the highly specialised inspector.

The Health and Safety Executive (HSE) is a body corporate under section 10(1) and consists of three persons "one to be appointed by the Commission with the approval of the Secretary of State to be Director of the Executive, and the others to be appointed by the Commission with the approval of the Secretary of State after consultation with the Director" (s.10(5)). The detailed provisions governing the Executive and its relations with the Commission are contained in Schedule 2 to the Act. It would seem that the Executive can operate when there are vacancies in its membership,[30] although, clearly, it could not operate indefinitely with, say, one member.

The HSE is now a complex organisation with 4,220 staff in post on April 1, 1980, after a recruitment ban imposed in deference to the Government's public expenditure cuts.[31] It now comprises the following inspectorates: factories, mines and quarries, nuclear installations, alkali and clean air, agriculture, and explosives together with supporting services. The supporting services consist of the Research and Laboratory Services, the Hazardous Installations Group, the Resources and Planning Division, and the Safety Policy Division. The Heads of the services together with the Director General, the Deputy Director General and the Director of Medical Services (discussed below) constitute the Management Board. This Board considers major policy questions and co-ordinates the work of the different divisions. The oldest of the inspectorates—the factories inspectorate—is also the most numerous, exceeding the total numbers of the other inspectorates.[32] Plans[33] had begun, before Royal Assent, to re-organise the factory inspectorate and in May 1975 the final proposals were announced.[34] The operation was completed in January 1977. Instead of one hundred and twenty-six regional and district offices with, on average, five inspectors per district office, twenty-one area offices were set up with up to thirty inspectors in each office. In addition, twenty-three smaller satellite or local offices were established to reduce inspectors' travelling time. Eighteen of the area offices are

[30] See Parl. Deb. H.C., November 4, 1976, col. 66 (Written Answers).

[31] HSC Report, 1979–80, Appendix 5.

[32] Factories Inspectorate—977; Explosives Inspectorate—15; Agriculture—188; Mines and Quarries—115; Nuclear Installations—106; and Alkali and Clean Air—48 (Parl. Deb. H.C., November 15, 1979, col. 731, Written Answers).

[33] See proposals for the re-organisation of H.M. Factory Inspectorate Report No. 1 (Revised), Health and Safety Commission Planning Unit, 1974.

[34] Industrial Relations Review and Report, vol. 114, October 1975, pp. 12–13. Department of Employment Gazette, October 1975. W. Simpson, Health and Safety—an Appraisal, Managerial Law VI, 27–33.

headed by an Area Director, two by Senior Area Directors and one by a Deputy Chief Inspector of Factories.

Each area office includes single industry groups of inspectors for those industries substantially represented in the area, a number of multi-industry groups, a construction industry team and a "services" team dealing with workpeople affected by health and safety legislation for the first time under the 1974 Act (the so-called "new entrants").

In addition, each area office includes a small group of inspectors who are responsible for co-ordinating the activities of the Factory Inspectorate nationally for a particular industry. Each national industry group provides a centre for the collection of data about practices, precautions and standards within a particular industry and, as such, a central forum in HSE for the analysis and discussion of the health and safety problems of the industry and the impact of broad proposals of the Commission and the Executive. The group can develop contacts with bodies representing interests in that industry—management, trade unions, suppliers of equipment and professional organisations—and is in a position to pinpoint health and safety problems in the industry so that health and safety performance in that industry can be improved. Particular problems in need of research can be identified and relative priorities can be established. Consistency of enforcement practice within the industry and, most important of all, stimulation of positive and constructive initiatives by those concerned in the industry are other advantages which national industry groups may bring in their train.

National industry groups have been formed for the co-ordination of work in the following twenty-one areas of industrial activity:—breweries, ceramics, chemicals, construction, docks, electricity, food, footwear, foundries, engineering, higher education, hospitals, paper, plastics, printing, research establishments, rubber, shipbuilding, steel, textiles (wool and cotton) wire and rope-making and woodworking.

Additionally, seven field consultant groups (FCG's) are now operational in Birmingham, Manchester, Edinburgh, Cardiff, Hitchin, Leeds and East Grinstead and are staffed by multi-disciplinary teams of scientists and engineers which provide a comprehensive technical and scientific service for factory inspectors in the field. Each has a base laboratory and a mobile laboratory equipped with the most modern instruments to facilitate the accurate measurement and analysis of toxic vapours and dangerous dusts which may be present in working environments, together with equipment for determining the composition of many chemicals found in industry.

The Factory Inspectorate itself (which now includes the

Explosives Inspectorate[35]) is responsible for the enforcement of the 1974 Act in relation to some 14 million people at work, located in some 500,000–600,000 premises. The combined strength of these two inspectorates on October 1, 1979 was 992.[36]

EMPLOYMENT MEDICAL ADVISORY SERVICE

Part II of the Act provides a new code governing the function of the Employment Medical Advisory Service. The service was established by the Employment Medical Advisory Services Act 1972[37] and started work on February 1st 1973. The system of appointed factory doctors originates from the earliest factory legislation. The Factories Act 1833 required a child to be examined by a surgeon or physician of the place or neighbourhood of the child's residence and by the Factories Act 1844 "certifying surgeons" were appointed who were later, in 1948, to be termed "appointed factory doctors."

The service is maintained in order to inform and adequately advise the Secretary of State for Employment, the Health and Safety Commission, the Manpower Services Commission and others concerned with the health of employed persons or persons seeking a training, on matters of which they ought to take cognisance, concerning the safeguarding and improvement of the health of their persons.[38] Further, it must give to employed persons and protected employees advice on health in relation to their employment and carry out other purposes of the Secretary of State's functions relating to employment.[39]

The service is now fully integrated into the Health and Safety Executive and its staff are accommodated in each of the twenty-one area offices. The service is headed by a Director of Medical Services and has within its staff specialist advisers in toxicology, respiratory diseases, pathology research and medical aspects of rehabilitation. The service numbers less than three hundred doctors (Employment Medical Advisers), nurses (Employment Nursing Advisers) and scientific support staff. Notwithstanding its small number of staff, EMAS sees it as an important part of its function to encourage others to pursue new areas of study by commissioning work in areas of growing

[35] H.M. Chief Inspector of Explosives works under the general direction of the Chief Inspector of Factories and provides a consultant service somewhat similar to that provided by H.M. Chemical Inspectors of Factories. The relationship between H.M. Explosive Inspectors and H.M. Chemical Inspectors of Factories has been rationalised.

[36] Parl. Deb. H.C., Issue No. 1151, November 15, 1979, (Written Answers) col. 731.

[37] Implementing the recommendations of The Appointed Factory Doctor Service. Report by a Sub-Committee of the Industrial Health Advisory Committee Ministry of Labour H.M.S.O. 1966.

[38] Section 55, Health and Safety at Work etc. Act 1974.

[39] *Ibid.*

concern. It has been estimated that appointed doctors conduct some 90,000 examinations annually. A further 10,000 are conducted by employment medical advisers.[40]

EMAS has by statute[41] the following functions:

To provide advice to the inspectorates on occupational health aspects of Regulations and Approved Codes of Practice; to carry out regular examinations of persons employed on known hazardous operations; to conduct medical examinations, investigations and surveys; to advise the Health and Safety Executive, employers, trade unions and others on the occupational aspects of poisonous substances; to investigate immunological disorders, physical hazards, dust and mental stress; to set standards of exposure for processes or substances which may harm health; to carry out research into occupational health; to provide advice on the provision of occupational medical, nursing and first aid services and to provide advice on the medical aspects of rehabilitation and training for and placement in employment.

EMAS has assumed an important rôle in relation to offshore installations. The Secretary of State for Energy, with the advice of the Director of Medical Services, has now approved a staff of over a hundred doctors to carry out the annual medical examination of divers, an examination which is required under both the Offshore Installations (Diving Operations) Regulations 1974[42] and the Submarines Pipelines (Diving Operations) Regulations 1976.[43] These doctors are also approved by the Secretary of State for Trade for the purposes of the Merchant Shipping (Diving Operations) Regulations 1975.[44]

The new Service provides a focal point for a comprehensive system of occupational medicine dealing with the wider issues of health such as stress and noise in addition to industrial diseases *simpliciter*.

Accident Prevention Advisory Unit

Originally part of the Factory Inspectorate the Accident Prevention Advisory Unit (APAU) examines the standards of occupational safety throughout industry and provides advice on the promotion of high safety standards within enterprises. The APAU is led by its Director and a small multidisciplinary team of Inspectors and is based in Preston, Lancashire.

[40] Parl. Deb. H.C., June 8, 1981, col. 233.
[41] s. 55, Health and Safety at Work etc. Act 1974.
[42] S.I. 1974 No. 1229.
[43] S.I. 1976 No. 923.
[44] S.I. 1975 No. 116.

The Executive had been established for only twelve months when its composition was vigorously attacked by a Royal Commission.[45] The Royal Commission noted:

> "that the terms of reference of the Robens Committee specifically excluded considerations of the question of general environmental pollution and that the Robens Committee's concern was with the health and safety of people at work and with the protection of the public from direct hazards arising from the workplace."

The Committee, said the Royal Commission, failed to see:

> "the implications of the transfer of the Alkali Inspectorate in terms of pollution control in its wider sense," and further "... it was an oversimplification to argue that common control arrangements should apply, for this requirement ignores great differences in the nature and scope of the interests of the two inspectorates."[46]

The Royal Commission concluded that:

> "the incorporation of the Alkali Inspectorate in the Health and Safety organization is potentially damaging to the interests of the environment"[47] and recommended "that the Alkali Inspectorate should be removed from the Health and Safety Executive forthwith and returned to the direct control of the Department of the Environment."[48]

These criticisms rightly continue to be rejected.

Problems of pollution caused by malfunctioning or inefficient factory plant can be effectively dealt with by specialist Inspectors[49] who principally visit to advise on safety and health. If these same factories were to be separately visited by a new Pollution Inspectorate there is no evidence to suggest that this duplication of very scarce and expensively trained manpower would produce a significant reduction in pollution.

In addition, the Alkali and Clean Air Inspectorate who would form the nucleus of such an inspectorate has in the past been strongly criticized for its weak enforcement of policy.[50] They have shown an extreme reluctance to take action through the

[45] Fifth Report of the Royal Commission on Environmental Pollution, Cmnd. 6471, H.M.S.O. 1976.
[46] *Ibid.*, para. 248.
[47] *Ibid.*, para. 257.
[48] *Ibid.*, para. 260.
[49] See p.13. In the case of very difficult technical problems they should be able to call upon their own expert group, that is the Alkali and Clean Air Inspectorate.
[50] For example, Frankel, M., "The Alkali Inspectorate—the control of Industrial Air Pollution." *Social Audit* Vol. 4, Spring 1974. "The Right to Know: An Investigation into Secrecy," *The Listener,* Vol. 91, 1974, pp. 559–562.

Courts.[51] There are now strong reasons to believe that the inclusion of the Alkali and Clean Air Inspectorate within the Health and Safety Executive is correcting this anomaly.

Early in 1981 new proposals for the re-organisation of the inspectorates were circulated internally. These included a proposal to introduce an Inspector General to be responsible for all Chief Inspectors, a proposal to remove quarries from the remit of the Mines and Quarries Inspectorate and a proposal to interchange staff between the inspectorates.[52] In April 1980, the Executive had 4,200 staff[53] and an annual net expenditure for the year ending March 31, 1981 of £69,824,000.[54]

LOCAL AUTHORITIES[55]

The Health and Safety at Work etc. Act 1974, section 18(2), provides that the Secretary of State may by regulations make local authorities responsible for the enforcement of statutory provisions.[56] The Robens Committee[57] argued that there was no feasible alternative to some sharing of responsibilities between central and local government. Its members believed that there were considerable intrinsic advantages in local authority inspection[58] and that many of the criticisms levelled against it would be reduced as a result of the re-organization of the structure of local government.[59]

It therefore recommended that local authorities should have a greater share of the work in this field, but that increased co-ordination and integration with the national Authority was

[51] Between 1920–1972 the Alkali and Clean Air Inspectorates brought fourteen successful prosecutions against registered works: 104th Annual Report on Alkali etc. Works 1967, H.M.S.O. 1968; 108th Annual Report on Alkali etc. Works, 1971, H.M.S.O. 1972.

[52] See Health and Safety Information Bulletin, May 1981.

[53] Report of the Health and Safety Commission, 1980–1981, H.M.S.O., 1981.

[54] *Ibid.*

[55] For the definition of local authorities, see Health and Safety at Work etc. Act 1974, ss. 53 and 54.

[56] Section 18(3) provides that any provision made under this sub-section shall have effect subject to any provision made by health and safety regulations in pursuance of Section 15(3)(*c*). Section 18(4) provides that it shall be the duty of every local authority to make adequate arrangements for the enforcement within their area of those provisions for which they are made responsible. The duties imposed on local authorities must be exercised in accordance with any such guidance as the Commission may give them. Section 45 provides for powers in the event of the failure of a local authority adequately to perform its enforcement function.

[57] Report of the Committee on Safety and Health at Work, Cmnd. 5034, H.M.S.O., 1972, p.73, para. 237.

[58] See the Report of the Committee of Inquiry on Health, Welfare and Safety in Non-Industrial Employment, Cmnd. 7664, H.M.S.O., 1949.

[59] There are currently 471 local authorities in England, Scotland and Wales empowered to enforce the Health and Safety at Work etc. Act 1974. See The Health and Safety (Enforcing Authority) Regulations 1977. (S.I. 1977 No. 746).

required.[60] The Committee thought it self-evident that the broad division of responsibility between central and local authorities should be influenced by the nature of the safety and health problems arising in different types of employment, by the nature of available expertise and by the need to avoid multiple inspection. It concluded that the central inspectorate should be responsible for industrial employment and that local authorities should concentrate on non-industrial employment.[61]

This division was made because the Committee believed that the central inspectorate possessed more expertise in industrial conditions and hazards whilst the local authority inspectorate were experienced in dealing with health and amenity. It argued that a broad criterion needed to be applied to reduce the number of cases where a particular establishment was inspected both by central and local authority inspectors.

Certain factors were, however, to prevent these principles from being adopted in full. Economic pressures on local authorities meant that funds were not available for the training of the very large number of Inspectors who were appointed by local authorities to enforce the Health and Safety at Work etc. Act 1974 under section 19[62] and, for the same reason, local authorities were unable to take on board many of the new entrants whose activities properly fell in the non-industrial sector.[63] Second, the activities of businesses do not neatly divide into industrial and non-industrial categories. Finally, other overriding issues had to be considered such as a Government preference[64] for its premises to be inspected by Civil Servants bound by the Official Secrets Acts. In that way it can ensure that state security is less likely to be breached.

The overall effect was to widen the inspectorial rôle of environmental health officers (formerly public health inspectors) in relation to health and safety matters, but *not* to change markedly the numbers of premises for which they were responsible.

[60] See the Report of the Committee on Safety and Health at Work, Cmnd. 5034, H.M.S.O., 1972, p.75, para. 243. In order to increase co-ordination and integration the Committee recommended that the National Authority be given powers of supervision over local authority inspection and that the Manager of each Area Office of the National Authority should be responsible for co-ordination, without prejudice to the statutory independence of local authorities, to ensure that a common set of standards was applied.

[61] See the Report of the Committee on Safety and Health at Work, Cmnd. 5034, H.M.S.O., 1972, p.76, para. 246. Special arrangements would need to be made for Crown premises where security considerations arose, and at hospitals and secondary schools where the range and type of hazards require special expertise.

[62] By July 19, 1977, 5,046 inspectors had been appointed by local authorities and of those 523 spent at least thirty hours per week enforcing the 1974 Act and its relevant statutory provisions; see Parl. Deb. H.C., July 19, 1977, col. 525.

[63] In July 1976 it was announced that the Government had decided not to make additional resources available to local authorities for health and safety enforcement.

[64] Principally the Treasury Solicitor.

The Health and Safety (Enforcing Authority) Regulations 1977[65] were laid before Parliament on May 6, 1977 and came into operation on June 1, 1977. Under these regulations, local authority inspectors lost their right of entry and inspection under health and safety legislation in relation to factories.[66] They gained the right, however, to enforce the general duties of the Health and Safety at Work etc. Act 1974 and any of the relevant statutory provisions[67] listed in Schedule 1 of that Act in those premises in which the main activity is listed in Schedule 1 of the 1977 Regulations.

Regulation 3 of the Health and Safety (Enforcing Authority) Regulations 1977 provides that "where the main activity carried on in any premises is specified in Schedule 1 to these Regulations, then subject to (certain exceptions) the local authority for the area in which those premises are situated shall be the enforcing authority for the relevant statutory provisions in relation to those premises and to any activity carried on in them."

The main activities listed in Schedule 1 are as follows:

1. The sale or storage of goods for retail or wholesale distribution *other than:*—(our emphasis)
   (a) on premises controlled or occupied by a railway undertaking;
   (b) in warehouses or other premises controlled or occupied by the owners, trustees or conservators of a dock, wharf or quay;
   (c) at container depots;
   (d) water and sewage and their by-products;
   (e) natural and town gas;
   (f) solid fuel or other minerals at any mine or quarry or at premises controlled from a mine or quarry;
   (g) petroleum spirit in premises where motor vehicles are maintained or repaired by way of trade;
   (h) wholesale distribution of flammable, toxic, oxidizing, corrosive or explosive substances or petroleum spirit.
2. Office activities.
3. Catering Services.
4. The provision of residential accommodation.

[65] The Health and Safety (Enforcing Authority) Regulations 1977, S.I. 1977 No. 746.

[66] Prior to the coming into force of these regulations, local authorities were responsible for the inspection of sanitary conveniences in all factories and for the inspection of sanitary conveniences, cleanliness, overcrowding, temperature, ventilation and the drainage of floors in factories where mechanical power was not used.

[67] Except those provisions where another inspector has been specifically named as the enforcing officer, for example: section 69 of the Factories Act 1961. For the list of existing relevant statutory provisions, see Schedule 1 to the Health and Safety at Work etc. Act 1974.

5. Consumer services provided in shop premises except dry cleaning or radio and television repairs.
6. Dry cleaning in coin-operated units in launderettes and similar premises.

In the majority of the premises listed local authorities previously inspected under the Offices Shops and Railway Premises Act 1963. Following the coming into force of the Health and Safety (Enforcing Authority) Regulations 1977 local authorities were made responsible for all the health and safety legislation which applied to those premises.[68]

For example, a large office block may contain, in addition to offices, a lift and lift rooms, a boiler house, a first-aid room, a staff canteen, printing machinery and guillotines, waste paper-baling machines, a boiler house and an underground car park. In such circumstances the Health and Safety at Work etc. Act 1974, the Offices Shops and Railway Premises Act 1963 and the Factories Act 1961 together with regulations made under those statutes will regulate conditions in those premises and will be enforced by the local authority in whose area the office block is situated.

The term "main activity" is nowhere defined in the regulations, but probably means the main purpose of employment of persons in the premises. This need not be the activity upon which greater numbers are employed or upon which greater space is utilised or which includes the greater hazard.[69]

Where the coming into force of the Health and Safety (Enforcing Authority) Regulations 1977 signalled the transfer of premises from the Health and Safety Executive to local authorities employers were notified.

The regulations were made with three important principles in mind. Firstly the allocation of premises would be made according to the main activity carried on. Secondly there would be no dual inspection, *i.e.* an attempt would be made to avoid visits to premises by both Health and Safety Executive inspectors and local authority inspectors. Thirdly there would be no self inspection, that is neither the Health and Safety Executive nor the local authority should inspect its own premises.

In some cases it has not been possible to reconcile these principles; for example some premises, even though they fall within Schedule 1, have not been transferred to local authorities because to have done so would have violated the principle of self inspection.[70] In addition Parliament has decided that premises,

[68] Subject to the proviso in note 67 *supra*.
[69] See LAAIC/B/2/2 Guidance published by the Health and Safety Commission for local authorities under section 18(4) Health and Safety at Work etc. Act 1974.
[70] See Regulation 4, Health and Safety Enforcing Authority Regulations 1977 op. cit.

and any activities carried on in them, which are occupied by or controlled by[71] County Councils, the Greater London Council and Regional Councils in Scotland, a police authority within the meaning of section 62 of the Police Act 1964 or the Receiver for the Metropolitan Police District, a fire authority within the meaning of section 43(1) of the Fire Precautions Act 1971, the United Kingdom Atomic Energy Authority and the Crown,[72] shall be inspected by the Health and Safety Executive.

In addition the principle of no dual inspection has suffered somewhat in that even though a local authority may be responsible for premises by virtue of regulation 3 and Schedule 1, the Health and Safety Executive is the enforcing authority for the following activities carried on in them:

(a) construction work[73] carried out by a person whose main activity is one for which the Executive is responsible for the enforcement of the relevant statutory provisions;
(b) the installation, maintenance and repair of gas, water or electricity systems where this is carried out by a person whose main activity is one for which the Executive is responsible for the enforcement of the relevant statutory provisions;
(c) the construction, installation, maintenance and repair of telecommunication systems;

Thus construction work carried out by construction firms in the private sector or the government service will be inspected by the Health and Safety Executive wherever that work is carried out in England, Scotland or Wales, but construction work carried out in a department store by employees of that store will be inspected by environmental health officers from the local authority.

Regulation 4(3) has been inserted to ensure that the licensing function and responsibility for ensuring the adequacy of fire precautions remains with the Greater London Council, county councils and regional councils in Scotland.

In London the Greater London Council has responsibility for enforcing the Offices Shops and Railway Premises Act 1963 in relation to offices and shops in music halls, cinemas and theatres.[74] It should however be noted that this responsibility does not extend to other employees at those premises such as

[71] Within the meaning of Health and Safety at Work etc. Act 1974, s.4.
[72] Except those occupied or controlled by the Health and Safety Executive.
[73] Construction work means a "building operation" and a work of engineering construction within the meanings assigned to those expressions by section 176(1) of the Factories Act 1961.
[74] See Offices, Shops and Railway Premises Act 1963, s. 52(5). All the powers of inspectors provided in the 1974 Act are also available to inspectors appointed by the Greater London Council.

usherettes, cleaners or projection staff[75] but the Health and Safety Executive is responsible for enforcing the Health and Safety at Work etc. Act 1974 in relation to the whole of those premises.

Regulation 5 provides for arrangements which enable responsibility for enforcement to be transferred from one enforcing authority to another by agreement. This may take place locally providing there is agreement between the enforcing authorities concerned or nationally at the behest of the Health and Safety Commission. Persons affected by such a transfer are entitled to be notified and may appeal against any such transfer to the Health and Safety Commission.[76] In relation to Crown premises the arrangements for transfer differ. Again, transfer may take place by the agreement but in this case the parties to that agreement include the Health and Safety Executive, the local authority concerned and the relevant Government Department. Any agreement reached can be cancelled by the three parties acting in concert or by the Health and Safety Commission.[77]

Regulation 6 provides for circumstances in which there is uncertainty as to whether the Health and Safety Executive or a local authority is responsible for the enforcement of the Act or its relevant statutory provisions in relation to particular premises or activities.

Regulation 6(1) provides:

> "For the purpose of removing uncertainty in any particular case as to what are their respective responsibilities by virtue of Regulations made under section 18(2) of the 1974 Act either the Executive or the local authority may apply to the Health and Safety Commission and where the Commission considers that there is uncertainty it shall, after considering the circumstances and any views which may have been expressed to them by either enforcing authority or by persons affected assign the responsibility to whichever authority it considers appropriate."

Such a case arose at Waverley Vintners Ltd., City Road, Newcastle-upon-Tyne. Early in 1979 a lift engineer, employed by a private contractor, was killed whilst working on a lift there. The accident was investigated both by the Executive and the local authority since some doubt arose as to which was the

[75] These are new entrants and are protected solely by the General Duties of the Health and Safety at Work etc. Act 1974.

[76] For example, employers, safety representatives, etc.

[77] In *Central Tyre Co. (South Side) Ltd.* v. *Ralph (Warwick District Council)* Birmingham Industrial Tribunal, September 15, 1978, HS 17863/78, where a local authority issued an improvement notice in relation to premises in relation to which it was not the enforcing authority, an Industrial Tribunal cancelled the notice.

enforcing authority. The premises were being used as a warehouse and the main activity therefore fell in Schedule 1 of the 1977 Regulations, but there was uncertainty as to whether the work being carried out on the lift fell within the construction work exception set out in Regulation 4(2)(*a*).[78] In this case the Health and Safety Commission considered the views of the Health and Safety Executive, the local authority, and the employer and the employees at the workplace. They also considered such factors as the premises' location, whether there were premises which were operated by the same employer in the immediate vicinity and the operational capability of the enforcing authority. The Commission decided to allocate responsibility to the local authority.

Regulation 6(2) provides that when responsibility for enforcement has been assigned the Commission must notify the authorities concerned and those persons affected by it.

The Health and Safety (Enforcing Authority)(Amendment) Regulations 1980[79] came into operation on December 29, 1980. They ensure that the Health and Safety Executive is responsible for enforcement in shop premises[80] in which the main activity is motor vehicle repair or maintenance.

Liaison at national level between the Health and Safety Executive and local authorities is provided by the Health and Safety Executive Local Authority Enforcement Liaison Committee (HELA). This Committee provides a forum for discussion on issues relating to local authority enforcement. The Health and Safety Executive is represented by a number of its senior staff whilst local authorities are represented by their associations. This Committee is invariably consulted before guidance memoranda are issued to local authorities under section 18(4) of the Health and Safety at Work etc. Act 1974. At local level, a principal inspector of factories, nominated by each area office of the Executive, acts as enforcement liaison officer and advises local authorities on enforcement matters.[81]

Fire

On January 1, 1977 fire authorities, or, in the case of Crown premises, H.M. Inspectors of Fire Services,[82] were given prime responsibility for enforcing general fire precautions in places of work subject to the Factories Acts and the Offices Shops and

[78] See page 57 *supra*.
[79] S.I. 1980 No. 744.
[80] As defined by the Health and Safety (Enforcing Authority) Regulations 1977, S.I. 1977 No. 746.
[81] Health and Safety Commission Report 1977–1978, H.M.S.O., 1979.
[82] Of the Home Office and Scottish Home and Health Departments.

Railway Premises Act 1963, in addition to places of work already their responsibility under the Fire Precautions Act 1971.

Section 17 of the Fire Precautions Act 1971[83] requires the fire authority to consult the enforcing authority before requiring alterations to buildings. Equally, enforcement officers under the 1974 Act are required to consult the fire authority where inadequate fire precautions are noted, where flammable substances are being used or stored, and where the service of a prohibition or an improvement notice is being considered which might affect fire precautions.[84]

Failure of a local authority to perform its enforcement functions

Section 45 of the Health and Safety at Work etc. Act 1974 provides that where the Health and Safety Commission is of the opinion that a local authority may have failed to perform any of its enforcement functions, the Commission may make a report to the Secretary of State.[85] The Secretary of State may, following such a report, order a local inquiry to be held.[86] If the Secretary of State is satisfied, following such an inquiry, that the local authority has defaulted, he may declare as such by order. He may, then, request the authority to perform such of their enforcement functions as are specified in such a manner as he specifies and he may also specify the time or times within which those functions are to be performed by the authority. If the defaulting authority fails to comply with such a direction, the Secretary of State may, instead of enforcing the order by mandamus, transfer such duties as he thinks fit to the Executive. The Executive will in such circumstances be entitled to charge such expenses as it certifies were incurred by it in performing the functions transferred. The Secretary of State is also vested with a further power to vary or revoke any order made in pursuance of this section.

[83] As amended by Health and Safety at Work etc. Act 1974, s.78.
[84] See Health and Safety at Work etc. Act 1974, s.23(4).
[85] *i.e.* the Secretary of State for Employment.
[86] See the Local Government Act 1972, s.250.

CHAPTER 3

THE GENERAL DUTIES

PRELIMINARY

Part I of the Act is designed to have effect with certain objectives in mind. These objectives provide a backdrop for the interpretation and application of the general duties discussed below. The four objectives set out in section 1(1) relate to the health, safety and welfare of persons at work, the health and safety of other persons who might be affected by the activities of the foregoing group, control of explosive and highly inflammable or other dangerous substances and control of emissions into the atmosphere of noxious or offensive substances.

Section 1(2) permits the existing protective Acts (as specified in Schedule 1) with their accompanying cohorts of regulations, orders and other instruments, to be "progressively replaced" by a system of regulations and approved codes of practice operating in combination with the other provisions of Part I and designed "to maintain or improve the standards" established by or under those enactments. The reference to the maintenance and improvement of standards was added to the Bill by the Government to allay the fears that the old standards might be relaxed in the new laws.[1] As we have seen, many of the duties in the "old laws" were (and are) absolute or strict; and critics who are worried about the stringency of new laws feel that this subsection should have been tightened up to prevent any dilution of the old standards, *e.g.* by reducing an absolute duty to one of reasonable practicability.[2] The Government rejected the argument that the stringency of the former standards might in some way be weakened by the promulgation of new standards under a phrase as broad as one which refers to the maintenance and improvement of standards, stressing that the phrase in question was not "just a presentational embellishment."[1] To have acceded to the views of the critics would have put any new regulations and codes into a legal straitjacket.

One other preliminary consideration which might be mentioned here, is the duty which section 55 of the Sex Discrimination Act 1975 places upon the Equal Opportunities Commission

[1] Parl. Deb. H.C., June 18, 1974, col. 306.
[2] See p. 14 *supra*.

to keep under review, in consultation with HSC, the relevant statutory provisions of the 1974 Act in so far as they require men and women to be treated differently. Following an extensive review of the legislation dealing with hours and conditions of work for women, the Equal Opportunities Commission has come to the conclusion that much of the protective law for women is out of date and is in need of reform.[3] Their report has been referred by the Secretary of State for Employment to HSC.

GENERAL DUTIES

(a) *Basic overriding responsibilities and defences*

Since previous health and safety legislation was too fragmented, too specific and contained no unifying or co-ordinating theme,[4] the Committee recommended—

> "that the Act should begin by enunciating the basic and overriding responsibilities of employers and employees."[5]

This general statement is now enacted in sections 2–9 of the Health and Safety at Work etc. Act 1974 and provides, *inter alia*, for safe working environments, safe equipment, trained and competent personnel and adequate instruction and supervision. Employees are required to observe these safety and health provisions and to act with due care for themselves and for others.

The Robens Committee believed that this general statement would encourage employers and workpeople to take a less narrow view of their responsibilities, provide guidance to the Courts in their work of statutory interpretation and encourage inspectors to look at the work place as a whole rather than with particular details which had been made the subject of regulations.[6]

(b) *"So far as is reasonably practicable"*

An important qualifying phrase in the Health and Safety at Work etc. Act 1974 is the duty to take safety measures "so far as is reasonably practicable."[7] This qualification was bitterly

[3] For a useful summary, see DE Gazette, April 1979, p.331.
[4] Committee on Safety and Health at Work Cmnd. 5034 H.M.S.O. p.41 para. 128.
[5] *Ibid.* p.41 para. 129.
[6] *Ibid.* p.41 para. 130–131. This was a reference to the checklist approach which had been adopted by Inspectors in the past. The Inspector was encouraged to raise his standards and look at the workplace as a whole. For further discussion of the new approach see Chapter 4 The Enforcing Authorities and their Policies.
[7] This phrase occurs on some eighty different occasions in the Health and Safety at Work etc. Act 1974.

opposed in a Standing Committee of the House of Commons by a number of Labour members[8] who mistakenly thought that its inclusion would greatly weaken the effectiveness of the Act.[9] The standard of reasonable practicability allows the courts and tribunals to take into account the many different circumstances inherent in different industrial operations. It is flexible and fair and seeks to reduce the negative influence of strict liability which some of the provisions of the previous legislation have sought to impose, but fulfils its objective in catching the thoughtless and the indifferent.

The phrase is not defined in the Act but its meaning was judicially determined in the leading case of *Edwards* v. *National Coal Board*[10] in the Court of Appeal and subsequently approved in *Marshall* v. *Gotham*[11] in the House of Lords.

In *Edwards* v. *National Coal Board,* the Court considered the phrase in relation to its meaning in section 102(8) of the Coal Mines Act 1911 and, in the words of Asquith L.J. stated—

> "Reasonably practicable" is a narrower term than 'physically possible' and seems to me to imply that a computation must be made by the owner, in which the quantum of risk is placed on one scale, and the sacrifice involved in the measures necessary for averting the risk (whether in money, time or trouble) is placed in the other; and that if it be shown that there is a gross disproportion between them—the risk being insignificant in relation to the sacrifice—the defendants discharge the onus on them. Moreover, this computation falls to be made by the owner at a point of time anterior to the accident."[12]

The Court in *Marshall* v. *Gotham Ltd.* had to deal with the death of a miner killed by a fall of the mine-roof caused by the occurrence of an unusual geological fault termed "slickenside" which was not detectable by any known means before a fall. As Lord Reid put it—

> "The only way to make a roof secure against a slickenside fall appears to be to shore it up, and, as the presence of slickenside cannot be detected in advance, full protection

[8] Official Report Standing Committee A Health and Safety at Work etc. Bill 2nd May 1974 cols. 71 and 78.

[9] For a discussion on the economic effect of the phrase "so far as is reasonably practicable," see Goldring M. Analysis: Your Money or Your Life B.B.C. Radio 4. June 3rd 1976.

[10] *Edwards* v. *National Coal Board* [1949] 1 K.B. 704 at page 712.

[11] *Marshall* v. *Gotham* [1954] A.C. 360 at page 373.

[12] At p. 712. See also *Jenkins* v. *Allied Ironfounders* [1970] 1 W.L.R. 304, at p. 307. Much depends on the state of knowledge of the risks at the time immediately before the accident: see *Adsett B.K.* v. *K. Steelfounders and Engineers Ltd.* [1953] 2 All E.R. 320.

> against this danger would require that every roof under which men have to pass or to work should be shored up or timbered. There is evidence that this is never done in gypsum mines and that in this mine the cost of doing it would be so great as to make the carrying on of the mine impossible."[13]

Lord Reid distinguished between measures which it might be practicable to take and reasonably practicable measures, although in the case itself he seemed to accept that use of props might have engendered "a false sense of security" and, whilst possibly reducing the chance of a fatal injury, would not have eliminated it.

The Americans have experienced a similar problem with standards issued under their Occupational Health and Safety Act 1970. Thus, in promulgating standards dealing with toxic materials or harmful physical agents under certain sections of that Act, the Secretary of State is obliged to set the standard which most adequately ensures, to the extent feasible, on the basis of the best available evidence, that no employee will suffer material impairment of health or functional capacity.[14] The American concept of "feasibility" is similar to, although not necessarily identical with, our concept of "reasonable practicability," in that it involves a balancing of the risk and the cost of averting that risk. The "best available evidence" is not necessarily the same thing as the most conservative scientific estimate. Whilst it might be justifiable to place our first lunar explorers in an extended period of quarantine on their return from the moon, the Supreme Court has invalidated an American standard which opts for a "play safe" standard for airborne benzene, thereby indicating that a conservative shot-in-the-dark may not survive the balancing test referred to above.[15] In this country, the question of costs was raised in the Report on Non-Departmental Public Bodies by Sir Leo Pliatzky, published in January 1980, which recommended that appraisals of the costs involved in health and safety measures to industry should in future be published. This will be done in consultative documents and draft regulations to be submitted to the Secretary of State. The Accident Prevention Advisory Unit within HSE refers in its Report on "Effective Policies for Health and Safety," 1980 (para. 20) to attempts to deal with costs by means of cost-benefit analysis, but concludes that—"Unfortunately the

[13] *Marshall* v. *Gotham* [1954] A.C. 360 at p. 371.
[14] *Ind. Union Dept. AFL–CIO* v. *Hodgson* 499 F. 2d. 467.
[15] See *Ind. Union Dept. AFL–CIO* v. *American Petroleum Institute* 65 L Ed. 2d. (1980) discussed below at p. 121.

techniques of cost-benefit analysis are not yet adequate to provide answers to questions of policy save in limited applications where the input and derivations can be identified in isolation." The Unit points out that some benefits are difficult to quantify, such as the spin-off in the form of good labour relations and increased productivity. Likewise, the assessment of costs takes us beyond the company or firm in that impaired health and safety of workers can result in social costs, *e.g.* in a nationalised health service.

There has been a continuing debate concerning the differences, if any, between the test of reasonable practicability and common law negligence. Although specifically there would seem to be little difference between reasonable practicability and reasonable care, the balance of judicial authority inclines to the view that the former is the higher standard. Sometimes the Courts have said that the common law duty is "no higher than"[16] the statutory duty whilst more usually and specifically they have said that "the statutory duty is higher than the common law duty."[17] These analogies between statutory and common law standards have arisen from the practice of joining civil claims under both heads. In *Smith* v. *National Coal Board*, Lord Reid declined to accept the argument that a regulation dealing with coal mine sidings was intended to codify the whole common law duty of the employer for the sphere in which the regulation operated.[18] It is not helpful to apply tests used in the statutory context to a common law claim and vice versa. In *Chipchase* v. *British Titan Products Co. Ltd.*[19] the plaintiff, a painter employed at the defendant's factory, fell from some staging which was nine inches wide. The staging was six feet above the ground; had it been six feet six inches above the ground regulations would have applied and a staging not less than thirty-four inches wide would have been required. The Court of Appeal declined to interfere with the decision that the nine-inch plank was sufficient at common law and gave a timely warning against the application of statutory standards by analogy.

Comparisons are inappropriate for the HSWA because this is a criminal statute and the burden of proof is reversed. In the prosecution of an offence under section 2 of the Health and Safety at Work etc. Act 1974, for example, it is first for the prosecution to adduce the evidence of injury and second for the

[16] *Bramham* v. *J. Lyons & Co. Ltd.* [1962] 1 W.L.R. 1048, 1052.
[17] *Powley* v. *British Siddeley Engines Ltd.* [1966] 1 W.L.R. 729, 732; *Trott* v. *W.E. Smith* [1957] 1 W.L.R. 1154.
[18] [1967] 1 W.L.R. 871.
[19] [1956] 1 Q.B. 545. See also the judgments of Lord Reid and Lord Tucker in *Marshall* v. *Gotham* [1954] A.C. 360, 373, 376.

employer to show, on the balance of probabilities that it was not reasonably practicable for him to have done more.[20] The standard is not a fixed standard. In fact it can involve increased levels of expenditure for more affluent employers; they may be expected to achieve a higher standard than the minimum.[21] In one of the first reported cases, *Belhaven Brewery Company Ltd.* v. *McLean*,[22] there was no dispute about the fact that the appellants were in breach of their statutory duties with respect to access to transmission machinery and other dangerous parts of the company's plant. What was disputed was an Improvement Notice issued by the Inspector which required screens fitted with an interlocking device (which would automatically shut off the machinery when a door was opened), whereas the employers considered that a screen without such a device could be made and fitted much more cheaply by their own millwright and that this would provide adequate protection. The fitting of the device was estimated to cost £1,900 compared to the £200 which would be involved if the millwright were to construct a screen. Notwithstanding the cost of the device, the intelligence and integrity of the operators and the high degree of supervision, the tribunal came down in favour of the device after balancing the cost of the device against the risk inherent in the use of the simple screen. It is interesting that the tribunal considered that remedial action in the form of the screen (without the device) would involve continuing breaches of strict duties in the Factories Act 1961 (ss.13(1) and 14(1) and also of s.2(1) of the HSWA 1974 which uses the test of reasonable practicability). In this latter connection they applied the test formulated in *Edwards* v. *National Coal Board*[23] and concluded that the risk[24] attendant upon an omission to fit an interlocking device was not so insignificant in relation to the sacrifice involved in fitting such a device that it could be said that the cost of fitting an interlocking device fell out the bounds of what was reasonably practicable.

In *Associated Dairies Ltd.* v. *Hartley*[25] the phrase "so far as is reasonably practicable" was again considered and on this occasion, in a reserved decision, the improvement notice was cancelled. The case concerned the use of hydraulic trolley jacks for the handling and transport of articles in the warehouse area of the ASDA Discount Centre in Grimsby. Fully loaded these trucks weighed two thousand kilogrammes and the Tribunal

[20] See section 40 of the Health and Safety at Work etc. Act 1974.
[21] Health and Safety Executive Policy Statement.
[22] [1975] I.R.L.R.370.
[23] See note 10, *supra*.
[24] The type of risk envisaged by the Tribunal in this case.
[25] [1979] I.R.L.R. 171.

found that if a single wheel of such a fully loaded roller truck were to pass over the foot of an individual a very serious injury would result to feet unprotected by safety shoes. Associated Dairies Ltd. provided a facility whereby employees could purchase safety shoes at cost price (*i.e.* between £8–£15) which were of a suitable specification; but Hartley, the inspector, issued an improvement notice requiring the company to ensure that all employees involved in operating trolley jacks should have available free of charge, suitable safety shoes. The tribunal decided that the matter should be considered in relation to all the sixty-six ASDA Stores in the country[26] of which the Grimsby operation was just one. The tribunal argued that this was because what is reasonably practicable in Grimsby is reasonably practicable at the other sixty-six branches.[27] The cost of providing such protective footwear free of charge at the sixty-six ASDA Stores was calculated to be in the region of £20,000 in the first year and £10,000 in each succeeding year. The tribunal further found as a fact that between 1972–1977 there had been only one accident with a roller truck involving injuries to a foot in the sixty-six depots contained in the ASDA group, whereas during 1977 there were ten accidents involving roller trucks in those sixty-six stores.

The tribunal's attention was drawn to three cases: *Qualcast (Wolverhampton) Ltd.* v. *Haynes*,[28] *Brown* v. *National Coal Board*[29] and *Belhaven Brewery* v. *McLean*.[30] The Tribunal of its own volition had regard to *Marshall* v. *Gotham*.[31] The Tribunal derived from a consideration of these cases the following guidelines in deciding whether arrangements were reasonably practicable. Firstly the tribunal had to decide whether the inspector's requirement was practicable. Their unanimous finding was that it was. Second in the whole circumstances whether the requirement was reasonable. The tribunal decided that this requirement should not be determined having regard solely to the proportion which the risk to be apprehended bears to the sacrifice in money, time or trouble involved in meeting the risk, but that it is proper to consider whether the time, trouble and expense of the inspector's requirement is disproportionate to the risk involved to the employees if the Tribunal were to cancel the improvement notice or affirm it with modifications. The tribunal, in a majority decision, held that the expense involved in providing the protective footwear free of charge was disprop-

[26] It is doubted whether the tribunal was correct to do this.
[27] But the notice was addressed to the Grimsby operation.
[28] *Qualcast (Wolverhampton) Ltd.* v. *Haynes* 1959] A.C. 743, H.L.
[29] *Brown* v. *National Coal Board* [1962] A.C. 574, H.L.
[30] *Belhaven Brewery* v. *McLean* [1975] I.R.L.R. 370.
[31] *Marshall* v. *Gotham* [1954] A.C. 360.

ortionate to the risk involved to the employees in requiring the employer merely to ensure that the safety shoes are available to wear in accordance with the existing arrangements

The phrase "so far as is reasonably practicable" was also interpreted in *Bartlett* v. *Newble*.[32] In that case one of Her Majesty's Agricultural Inspectors was of the opinion that the use of an oast house near Maidstone, Kent would involve an imminent risk of serious personal injury. The prohibition notice directed that the oast should not be used until a schedule of remedial works had been carried out. The owners appealed arguing, *inter alia,* that it was "vital to [their] farming operation that [they] [had] the use of the oast to dry this year's crop of hops since there [was] no alternative means of drying them on the farm."

The oast building was situated on a farm which contained 48 acres of hops in full production some 600 to 650 cwt. of dried hops being produced annually. The season lasted three or four weeks during September. During that period two people would work in the building day and night. In addition two others would come in during the day. For the rest of the year the oast building was used for storage purposes.

The building was examined and found to be in a seriously unstable condition and in imminent danger of collapse. It was also not disputed that if it were to collapse whilst someone was inside that person would almost certainly be killed or seriously injured.

The appellants agreed it was necessary to take remedial measures but disputed whether this was necessary before that year's hop picking season and asked for twelve months to do the work.

The tribunal held that, where men's lives were at stake, they would not lightly hold that the taking of practicable steps was unreasonable and approved the prohibition notice.

"Reasonably practicable" also falls to be considered in relation to whether appeals against improvement notices and prohibition notices can be heard after the period for their submission has expired. In the Industrial Tribunals (Improvement and Prohibition Notices Appeals) Regulations 1974 (S.I. 1974 No. 1925), regulation 3 provides for rules of procedure in relation to appeals to a tribunal under section 24 of the principal Act against improvement and prohibition notices relating to matters in England and Wales. These rules of procedure are set out in a Schedule. Rule 2(2) provides:

[32] *Bartlett* v. *Newble* (1979) HS 18046/79 (unreported).

> "A tribunal may extend the time mentioned above where it is satisfied on an application made in writing to the Secretary of the Tribunals either before or after the expiration of that time that it is not or was not reasonably practicable for an appeal to be brought within that time."

It is clear, however, that this extension of time should not be given lightly. This discretion must be exercised by the full tribunal and not by the President or Chairman acting alone.[33] Some guidance in the interpretation of this provision is available from a consideration by the Court of Appeal of the Industrial Tribunals (Industrial Relations etc.) Regulations 1972 (S.I. 1972 No. 38), Sched., rule 2(1), which provides, so far as is material:

> "In relation to proceedings on complaints under section 106 of the Industrial Relations Act 1971 a tribunal shall not entertain such a complaint unless it is presented before the end of the period of four weeks beginning (*a*) In the case of a complaint relating to dismissal, with the effective date of termination . . . unless the tribunal is satisfied that in the circumstances it was not practicable for the complaint to be presented before the end of that period."

The interpretation of this provision (and its successor provision[34]) was aptly summarised by Stephenson L.J. in *Porter* v. *Bandridge.*[35] It was here explained that if the complainant does not know of his rights or of the time limit and ought not to have known of them because there was nothing to put him on inquiry he is not at fault and the industrial tribunal should be satisfied that it was not reasonably practicable. But if he is at fault, or he goes to solicitors who are at fault, in allowing the three months to go by, the industrial tribunal should not be satisfied. But ignorance of one's rights is not necessarily a good excuse—the tribunal must look at all the circumstances, what opportunities he had for finding out his rights and what explanation can be given for his ignorance of them.

[33] Rule 14(3) Schedule, The Industrial Tribunals (Improvement and Prohibition Notices Appeals) Regulations 1974 (S.I. 1974 No. 1925). For an examination of this provision, see *Fantarrow* v. *Leworthy* HS/21905/79 (unreported). See also *Caravan Parts (Supply) Ltd.* v. *Peacey* HS/11776/80 (unreported) and *Brew Brothers Ltd.* v. *Mallon* HS/17106/76/A (unreported).

[34] The Trade Union and Labour Relations Act 1974, Schedule 1 para. 21(4), now repealed by Employment Protection (Consolidation) Act 1978.

[35] *Porter* v. *Bandridge* [1978] 1 W.L.R. 1145.

General Duties of Employers and their Employees

THE PORTMANTEAU DUTY: SECTION 2(1)

In contrast to the previous emphasis upon premises, persons and processes, section 2(1) makes it the duty of *every* employer to ensure so far as is reasonably practicable, the health, safety and welfare at work of his employees. The result is to extend the coverage of health, safety and welfare protection to new categories of persons—the "new entrants" as they are termed. Hospital, schools and universities although not carried on for "gain" (as is required of processes in factories) are covered by the new duties. These new duties have added considerably to the burden of enforcement which falls upon the enforcing authorities. Responsibility for the new entrants has, to their embarrassment, been allocated to H.M. Factories Inspectorate, and attempts to share the burden with local authorities have, for the time being, been shelved.[36] As has already been indicated, these new duties constitute for many persons an "upper tier" of regulation additional to the existing tier of statutory regulation, *e.g.* the Factories Act and delegated legislation thereunder. For the new entrants, not previously covered by the "codes," the new duties will be the "only tier" (ignoring the fact that one cannot have a one-tier ranking!) In due course they will be brought within the coverage of new regulations. The duty of every employer to ensure so far as is reasonably practicable the health, safety and welfare at work of his employees requires that employers investigate complaints promptly and sensibly.[37]

The portmanteau duty is owed to "employees," namely those who work under contracts of employment or apprenticeship, provided that they are "at work." An employee is "at work" throughout the time that he is in the course of employment but not otherwise. The concept "course of employment" is a prolific source of litigation in connection with the law of tort and doubtless many of the tort decisions will be resorted to in the present context. One thing is clear, and that is that a person may be "at work" without actually doing work, *e.g.* preparing a machine, clearing away at the end of a shift, breaking for meals and so forth. Regulations may extend the meaning of "work" and

[36] See further *The New Institutional Structure*, p. 54 and footnote 63 above. Local authority associations, and the Association of Metropolitan Authorities in particular, have decided as a matter of principle not to accept new duties without a corresponding addition to the Rate Support Grant.

[37] See *British Aircraft Corporation* v. *Austin* [1978] I.R.L.R. 332 (in relation to ill-fitting safety goggles provided for an employee).

"at work" and may adapt any of the relevant statutory provisions.[38] In most cases, travelling to and from work will lie outside the protection of the section, although it is significant that I.L.O. Recommendation No. 121, Art. 5(*c*) would include accidents travelling to and from work and the place of residence.

In four of the five specific duties given in section 2(2), four make reference to the "provision" or "maintenance" of certain things. Where premises or permanent machinery, such as a lathe or power-press, are involved, there will be little doubt that these are provided. The primary meaning of the word "provide" is to "furnish" or "supply." Clearly, the term denotes that something must be available, *e.g.* goggles or protective footwear. As Romer L.J. put it in one case—"A butcher does not 'provide' or 'supply' his customer with meat if he leaves it at the roadside a mile away from the customer's house."[39] How far the duty extends beyond this is not entirely certain. For example, in *Bux* v. *Slough Metals Ltd.*,[40] goggles had been given to each die-caster in a foundry, with instructions that they must be worn at work. It was held that the employers had discharged their statutory duty to "provide" goggles in accordance with regulations, but were in breach of their common law duty to safeguard the employee's safety by not instructing employees in a reasonably firm manner that the goggles must be used followed by supervision of their use. The case illustrates the danger in borrowing civil standards from a damages suit for the purpose of arriving at standards in regulations.

The civil law gives us concepts such as causation linking breach of duty and resulting harm, hypothetical causation in which the breach of duty does not cause the harm because of the safety device or measure would not have been used by employees, self-evident negligence (*res ipsa loquitur*) in relation to a failure to inspect plant or equipment provided[41] or some act such as failure to inspect a tool which breaks the chain of causation (*novus actus interveniens*).[42] These civil law concepts, which are all concerned with injury and loss, have no part in the criminal law, in that a criminal does not in the present context require injury or loss as a necessary ingredient.

[38] See, for example, regulation 3 of the Health and Safety (Genetic Manipulation) Regulations 1978 (S.I. 1978 No. 752).
[39] *Norris* v. *Syndic Manufacturing Co.* [1952] 2 Q.B. 135, at p.146.
[40] [1974] 1 All E.R. 262.
[41] *Barkway* v. *South Wales Transport Co.* [1950] W.N. 95, H.L.
[42] *Taylor* v. *Rover Car Co.* [1966] 1 W.L.R. 1491.

THE FIVE ILLUSTRATIONS OF THE PORTMANTEAU DUTY

Without prejudice to the generality of the portmanteau duty, section 2(2) makes specific that duty in five examples which are themselves of fairly general application and all of which use the test of reasonable practicability.

(a) Plant and Systems of Work: Section 2(2)(a)

The employer is obliged to provide and maintain so far as is reasonably practicable plant and systems of work which are safe and without risks to health. The term "plant" is defined in section 53(1) and "includes any machinery, equipment, or appliance." The central notion is of something which is used for the purposes of the employer's business or enterprise such as the "fixtures, implements, machinery and apparatus used in carrying on any industrial process," (according to the O.E.D.). The term means something used in the industrial process and not the products of that process or the objects on which that process is carried out. A graphic example is provided by *Haigh* v. *Charles W. Ireland Ltd.*,[43] which turned upon a provision in the Factories Act 1961 forbidding the application of heat or cutting operations to any plant, tank or vessel which contains or has contained any explosive or inflammable substance, and which provision was held to be inapplicable to the use in a scrap metal yard of an oxyacetylene cutter (with disastrous results) to cut open a safe containing dynamite. The safe was not plant within the statutory meaning of that word. Similarly, the fencing provisions of the Factories Act 1961 do not apply to dangerous parts of machines which are themselves the product of the factory.[44] The distinction between fixed and circulating capital is axiomatic in accountancy theory (similar to, but not identical with, the distinction between capital and revenue expenditure in revenue law). The safe in the Haigh Case, above, might well have been "plant" in the business in which it was used (*e.g.* in a wages office), but it became the "stock-in-trade" or "raw material" of the scrap metal business. It does not follow that because a thing is not technically "plant" there is no statutory duty with regard to it; duties with regard to a safe system of work, safe articles and substances, or a safe place of work provide a safety net.

The employer's duty to provide and maintain safe systems of

[43] [1973] 3 All E.R. 1137.
[44] *Thurogood* v. *Van Den Berghs and Jurgens Ltd.* [1951] 2 K.B. 537.

work is redolent of the employer's common law duty of care as stated in the leading case of *Wilson and Clyde Coal Co.* v. *English*,[45] in which the employer's obligation to exercise reasonable care for the safety of his employees was classified into sub-categories *viz.* proper personnel and supervision, proper plant and equipment, and a safe system of work—a threefold classification which has been judicially criticised as providing no more than three instances of a single general duty of care.

Despite the warnings which we have given against borrowing too freely from the common law, the view that the general duties in sections 2 to 9 of the Act "are largely modelled on the common law duties of care"[46] has received judicial confirmation in *R.* v. *Swan Hunter Shipbuilders Ltd. and Telemeter Installations Ltd.*[47] (discussed below at p. 75), the first case to be tried by jury under the Act. Accordingly, it seems that the fulsome common law on "safe system of work" will be prayed in aid in the present statutory context. So far as the common law is concerned a "system" could include the following. First, and least controversial, it would include a permanent organisation of work which is of a settled and continuing nature, *e.g.* work on an assembly line. Second, work which varies from one project to another, *e.g.* construction work. Third, an isolated task which in the words of Lord Reid is "an operation . . . of a complicated or unusual character," such as removing a heavy cog wheel from a lorry.[48] The employer must take account of the inexperience, personal infirmities and known propensities of his employees, although all work to a varying extent involves judgment and discretion in the employee the exercise of which would fall outside the concept of a "system" for which the employer is responsible.[49]

The concept of "a system of work" breaks loose from the central notion in the codes that the statutory duties apply only within defined spatial limits, the factory, the shop, the office and so forth. An employer's obligation under s.2(2)(*a*) is not geographically limited in this way and comes to resemble the mobility of common law negligence in which employers have been held subject to a duty of care in respect of premises outside their control. *General Cleaning Contractors Ltd.* v. *Christmas*[50] concerned the common practice of window-cleaners who stand on window sills to clean the outsides of sash windows, the sash

[45] [1938] A.C. 79. See the clear account in Munkman, Employer's Liability, 9th edition, pp.123 *et seq.*
[46] Munkman, *Employer's Liability*, 9th ed., p. 206.
[47] [1981] I.R.L.R. 403.
[48] *Winter* v. *Cardiff R.D.C.* 1950 1 All E.R. 819, p.825; *Rees* v. *Cambrian Wagon Works Ltd.* (1946) 175 L.T. 220.
[49] *Hoover* v. *Mallon* (1978) I.D.S. Brief 131 p.16.
[50] [1953] A.C. 180.

on one occasion moved causing the window-cleaner to fall. Notwithstanding that the employees were working on the premises of customers of the employer, the latter was held liable for failing to instruct their employees in a safe system of cleaning sash windows from the outside and for failing to provide safety devices, such as wedges, to prevent sashes moving. The possibility of using alternative methods such as ladders, cradles or safety belts, or of holding that the operation in question was so inherently dangerous that it ought not to have been performed at all, were two possibilities which did not commend themselves to the House of Lords, Lord Tucker categorically denying that "it was reasonably practicable for the defendants to insist on being allowed to insert hooks (*for safety belts*) into the brickwork of the building"(*italics supplied*).[51] On the other hand, as the subsequent case of *Wilson* v. *Tyneside Window Cleaning Co.*[52] shows, there is an area of judgment and discretion left to the employee. In that case the employer was held not liable for negligence in failing to inspect premises or to warn experienced window-cleaners of the danger in pulling too hard on handles attached to window sashes. Distinguishing between premises of the employer and premises of a stranger, Pearce L.J., whilst accepting that the duty of care applies to both, considered that "as a matter of commonsense its performance and discharge will probably be vastly different in the two cases. The master's own premises are under his control: if they are dangerously in need of repair he can and must rectify the fault at once if he is to escape the censure of negligence. But if a master sends his plumber to mend a leak in a respectable private house, no one could hold him negligent for not visiting the house himself to see if the carpet in the hall creates a trap."[53] Reasonable practicability is a relative test and must depend upon the circumstances. The protean nature of the common law duty of the employer goes wider than the duty to provide a safe system in section 2(2)(*a*) so that, for instance, the act of the employer in sending a fifty-seven year-old employee on a long journey in an unheated van with a defective radiator was held to amount to negligence,[54] but it is doubtful if that same act would amount to an unsafe "system" under that subsection.

As with the common law duty, the statutory duty is not only to provide but also to maintain safe systems of work. The statutory duties in section 2 are owed by an employer to his employees.

[51] S.C. p.198.
[52] [1958] 2 Q.B. 110.
[53] S.C. at p.121.
[54] *Bradford* v. *Robinson Rentals Ltd.* [1967] 1 W.L.R. 37.

The Swan Hunter Case[55] shows that in discharging those duties the employer may have to undertake responsibilities to persons who are not his. The case arose out of a fire which killed eight men working on HMS Glasgow in the Swan Hunter shipyard on the River Tyne. The fire had been caused when a welder, without any negligence on his part, struck his arc with a welding torch thereby igniting the atmosphere which had become enriched with oxygen because B, a former employee of Swan Hunter's, but at the relevant time an employee of Telemeter Installations Ltd., had failed to turn off the oxygen supply in a confined space when he left work on the previous evening. Swan Hunter had taken such steps as were reasonably practicable to inform and train their *own* employees in the dangers concerning oxygen equipment and pleaded that their duty to provide a safe system of work under s.2(2)(*a*) could not be held to extend to the employees of the several contractors working on the ship (with whom there might not be contractual relations, as in the case of Telemeter Installations) and, also, that they were not obliged to instruct employees of other employers under s.2(2)(*c*) over whom they had no control. Both firms were however convicted by the Crown Court and lost their appeals against conviction in the Court of Appeal. Dunn L.J. agreed with the trial judge that the general duty in s.2(1) (the portmanteau duty as we have termed it) is strict and covers the following particular duties so that—"If the provision of a safe system of work for the benefit of his own employees involves information and instruction as to potential dangers being given to persons other than the employer's own employees, then the employer is under a duty to provide such information and instruction,"[56] adding that his protection lay in the words "so far as is reasonably practicable" where the onus lay on him to prove on a balance of probabilities that it was not reasonably practicable to ensure a safe system in the circumstances of the present case. It will be seen therefore that an employer's duty to his own employees may be broken when he fails to give information and instruction to the employees of another employer where both employers are engaged on some common project. The duties of employers and the self-employed to persons not in their employment (again modelled on the common law duties of care) are contained in section 3, which is discussed below.

[55] [1981] I.R.L.R. 403.
[56] S.C. at p. 407.

(b) *Articles and Substances: s.2(2)(b)*

The second application of the general duty of the employer relates to the arrangements which he must make to ensure, so far as is reasonably practicable, the safety and absence of risks to health in connection with the use, handling, storage and transport of articles and substances. The matter, as the Director-General of HSE has pointed out, is not one of principle but one of knowledge. Scientific and medical research (the latter depending increasingly on epidemiological techniques) add almost daily to the list of articles and substances which despite previous popular acceptance as harmless are seen to have dangerous properties. The Thalidomide Affair and research on the anti-contraceptive "pill" have had a chastening effect on too euphoric an acceptance of technical progress. A thing is dangerous whether or not the user knows of its dangerous propensities; but he cannot take measures against a risk of which he is unaware. An employer cannot be expected to take safety and health measures in respect of articles and substances not known to have harmful propensities. Thus the harmful effect of welding fumes in low concentrations over a prolonged period of exposure was not perceived in 1968 despite one speculative publication to the contrary. Consequently an employer was held not to be in breach of provisions in the Factories Act 1961 requiring him to take measures against injury to health arising from fumes dust and impurities.[57] In any case, the intention of a provision as we have seen may be to protect against a possible type of injury, such as a duty to provide exhaust devices to prevent the inhalation of dust, a provision which would not extend to cover someone who contracts dermatitis by contact with the dust. Although it is a truism that one cannot take steps to "make safe" something not known to be "unsafe" there may be a sweeping up provision such as that in section 2(2)(*d*) which refers to "welfare at work" (see below). One of the factors which has caused public concern has been the seemingly slow response to control substances which are known to be dangerous; for example the slowness in reacting to the asbestos dangers at Hebden Bridge or to the use of vinyl rubber compounds as a raw material with several manufacturing applications. The enforcing authorities are now most anxious not to be seen to be closing stable doors after the horses have gone and have turned a great deal of attention and resources to identifying the dangerous propensities, particularly the capacity for long-term harm, of new substances.

[57] *Cartwright* v. *G.K.N. Sankey* (1972) 12 I.R.R. 453 (where *dicta* in *Ebbs* v. *James Whitson and Co. Ltd.* [1952] 2 Q.B. 877 were not followed).

Thus, as the result of a smallpox occurrence in 1978 at Birmingham University, in which a photographer in the University Medical School died, a committee appointed to report on the occurrence recommended legislation, which is now to be found in the Health and Safety (Dangerous Pathogens) Regulations 1981 (S.I. No. 1011), regulating the keeping, handling or transportation of listed pathogens such as smallpox virus or Lassa Fever virus. Birmingham University was unsuccessfully prosecuted as a result of that occurrence. The case serves as a reminder that there is no *res ipsa loquitur* doctrine in cases of this sort. It is tempting, but incorrect, to say that the occurrence must *of itself* demonstrate that reasonably practicable precautions were not taken.

The Health and Safety Commission uses Advisory Committees[58] to advise it in the work of identifying dangers and in the promulgation of standards and legislative measures to reduce those dangers to acceptable limits. The Advisory Committee on Toxic Substances reviews substances, such as carcinogenic agents, with a view to determining standards, standards which in the case of lead have resulted in Regulations[59] and a Code of Practice. Other Committees are the Advisory Committees on Dangerous Substances[60] and on Major Hazards. The Advisory Committee on Asbestos completed its work in 1979 with the publication of its third and final Report which, with the two earlier reports, has already led to a Code of Practice and a Guidance Note which supplement the Asbestos Regulations 1969 (S.I. No. 690) and will lead to further measures and standards.[61]

As in the case of negligence, the duties under section 2 may need to be related to the individual employee, as in *Page* v. *Freight Hire (Tank Haulage) Ltd.*,[62] where an employer refused to allow a divorced female lorry driver to transport a chemical notified to him by the manufacturers as potentially dangerous to women of child-bearing age. Her complaint of a detriment, contrary to the Sex Discrimination Act 1975, was dismissed, the E.A.T. putting the justification for the employer's decision on section 51(1) of the 1975 Act (which negatives discrimination in order to comply with a requirement in an Act passed before the 1975 Act) and not upon some general canon of "safety" which was too broad a rubric.

[58] See *The New Institutional Structure*, above p. 45.
[59] Control of Lead at Work Regulations 1980 (S.I. No. 1248) replacing previous regulations.
[60] See the Packaging and Labelling of Dangerous Substances (Amendment) Regulations 1981 (S.I. 792) which in listing some 800 dangerous chemicals implement an EEC Directive (79/370: OJ No. L88/79). Also the Hazardous Substances (Labelling of Road Tankers) Regulations 1978 (S.I. No. 1702).
[61] HSC Report 1979–80, paras. 40–45.
[62] [1981] I.C.R. 299.

(c) *Information, Instruction, Training and Supervision: Section 2(2)(c)*

Unless employees have proper information, training and supervision the best-laid plans of management are apt to fail, something which section 2(2)(*c*) recognises when it requires the employer to provide such information, instruction, training and supervision as is necessary to ensure, so far as is reasonably practicable, the health and safety at work of his employees. No longer is it possible to place health and safety behind a shroud of secrecy; the worker has a "right to know." It is significant that the chief complaint of workers in the Seveso Tragedy in Italy, when an escape of airborne chemical wrought tremendous damage to human life and to the local ecology, was the complete lack of information following the event. Soldiers in the second world war, unlike their predecessors in the 1914–18 war, were usually "put in the picture," as it was termed, because it was found that they responded better when they knew the risks and what they must do to reduce those risks. Failure to communicate proper information may lead to rumours and a "China Syndrome" just as the release of inchoate data may itself have a similar effect.

The fourfold duty under section 2(2)(*c*) with regard to information, instruction, training and supervision is part of a battery of requirements which have as their objective the involvement of well-informed employees and others in activities at work. The overall pattern can be seen from the following:

(i) Employer's duty to provide a safety policy (s.2(3)).[63]
(ii) Appointment of safety representatives and safety committees (s.2(4), (6) & (7)).[64]
(iii) Duty of HSC to provide information (s.11(2)(*b*) & (*c*))[65]
(iv) Duty of an inspector to provide information (s.28(8)).[66] Although the prime duty to provide information rests on employers, inspectors have, in the words of the HSC, "an important complementary role in providing information and advice from their own knowledge."[67]
(v) Regulations may be made requiring information on health and safety to be included in directors' report (s.79).[68]
(vi) Linked with (i)–(v) is the duty of the employer (not yet operative) to give to those who are not his employees

[63] See below, p. 83.
[64] See below, p. 84.
[65] See below, p. 178.
[66] See below, p. 183.
[67] HSC Report, 1978–79, para. 7: HSC Newsletter No. 6, June 1979.
[68] See p. 84, below.

information concerning the conduct of the undertaking which might affect their health and safety (s.3(3)).[69]

(vii) Duty of manufacturers of articles and substances for use at work to make available adequate information about their use, about relevant tests and about conditions under which they may safely be used[70] (HSC have issued a Consultative Paper in 1975 proposing ways in which this statutory duty could be complied with).

(viii) The Employment Medical Advisory Service (EMAS) has the function under the Act of keeping everyone concerned with the health of employed persons informed of matters about which they ought to know in order to safeguard and improve health.[71]

The information which is provided must be accurate and as comprehensive as possible. *Vacwell Engineering Ltd.* v. *BDH Chemicals Ltd.*[72] was a case in which suppliers of a dangerous chemical were held liable in contract and for the tort of negligence; but it is submitted that had the defendants been employers they would, as a company dealing in chemical agents, have been obliged to conduct adequate research into the scientific literature on a particular chemical where that research would have shown a risk of explosion if the chemical were to be brought into contact with water. In other words it is not merely the duty of the employer to communicate the knowledge which he has but also the knowledge which as an employer in that position he ought to have possessed. Proper and clear instructions must be given as to what the employee is required to do or not to do bearing in mind that workers performing repetitious routine tasks are not infrequently careless as is sometimes the case with young and inexperienced workers. The duties are *ad hominem* so that, for example, the information and instruction which must be given to employee whose command of English is weak must take account of their linguistic difficulty.

Section 2(2)(*c*) generalises duties in several existing Acts. The existing duties are set out in sections 135–137, inclusive, of the Mines and Quarries Act 1954, section 138 of the Factories Act 1961 and section 50 of the Offices, Shops and Railway Premises Act 1963, which are supplemented by subordinate regulations dealing with information and the posting of abstracts. The existing system has been criticised by HSC in a consultative document issued in 1975, which makes proposals for regulations

[69] See p. 87, below.
[70] s. 6 on which see below, p. 92.
[71] s. 55(1) of the 1974 Act.
[72] [1971] 1 Q.B. 88.

to amplify the nature of the duties under section 2(2)(*c*). The existing duties are criticised because (1) they are limited to the provision of information on the law (as opposed to advisory or informational material generally); (2) they rely heavily on the posting of placard copies of abstracts which are rarely read; (3) they are limited to the clientèles covered by the existing codes (*i.e.*, they would not apply to the "new entrants"); and (4) they become obsolete and need to be revised from time to time. In its proposals for new and comprehensive regulations on the provision of information to employees, the HSC suggests: notices giving details of arrangements for the provision of information; the provision of access by employees to publications issued by HSC or HSE relevant to health, safety and welfare at work; and arrangements for employees to consult statutes, regulations, orders, codes of practice or guidance on health and safety at work. One of the advantages of the proposals according to HSC would be that "it would establish the principle that compliance with section 2(2)(*c*) required the setting up of a proper information outlet (however modest) by every employer. This might be the basis on which other requirements could rest in the future."[73] Not surprisingly, the HSC proposals have not been taken up. It would seem that the sheer cost falling upon employers to provide an information service on this scale is the main obstacle to implementation of the proposals.

(d) Place of Work: Section 2(2)(d)

By virtue of s.2(2)(*d*), the general duty of the employer is specifically extended to include so far as is reasonably practicable a place of work under his control which is safe and without risks to health and the provision and maintenance of means of access and egress to and from such place of work which are safe and without such risks.

The concept of "control" is wider than the concept of ownership; an employer may well control premises which he does not own, *e.g.* where the premises are occupied under a mere licence. The Robens Committee recognised that the obligations which could be placed upon employers would depend upon the circumstances, referring to those employees who spend most of their time away from a fixed base. We have seen that the duty of an employer in relation to his employees who are window-cleaners must take account of his inability to insist on structural safety measures in the premises of his clients, *e.g.* driving safety hooks right through outside walls. Whilst death and serious

[73] *Ibid.*, para. 15.

injury capture the headlines, most people are injured by tripping on things or slipping on staircases or floors.

To some extent, the duty in section 2(2)(*d*) overlaps with existing provisions regulating the safety of the place of work, together with safe access and egress to and from such place. Examples are sections 28, 29 of the Factories Act 1961, section 16 of the Offices, Shops and Railway Premises Act 1963, and the several relevant provisions in the Mines and Quarries Act 1954. Many of these statutory provisions are well-illustrated by case-law and until such time as they are replaced by regulations it is likely that inspectors and prosecutors will prefer the devil they know. The duty in section 2(2)(*d*) will be the sole protection available to the "new entrants" until such time as regulations make special provision for them. Unlike the existing statutory provisions, the duty in section 2(2)(*d*) extends to "risks to health."

The meaning of "any place of work" is not entirely clear, particularly in view of the doubt as to the extent to which the Courts will resort when construing like expressions in the existing statutory provisions. The phrase may mean the actual place at which work is performed (similar to the phrase "at which any person has at any time to work," as used in section 29(1) of the Factories Act 1961), or the place of work as a whole, such as a factory and its curtilage. The latter meaning would make it difficult to accommodate the additional duty concerning access and egress which are likely to be outside the employer's control. If the former meaning is preferred it would still remain to be decided how far places such as the canteen, lavatory or first-aid room would be included. A break with the literalism condemned by the Robens Committee would permit the inclusion of such places either as places of work or means of access and egress to and from work. The duty is not absolute but tempered by the test of reasonable practicability. An absolute test would require the employer to close the factory temporarily inundated by floodwaters, as in *Latimer* v. *A.E.C.*,[74] or which is affected by some transient condition. The balancing of risk versus measures to avert the risk may allow the employer to keep his factory going notwithstanding the slippery state of the floor left by the flood.

It is necessary to distinguish between the duty of an employer to his employees to provide and maintain safe access and egress under section 2(2)(*d*) and the duty owed by controllers of

[74] [1953] A.C. 643 HL.

premises to provide safe access and egress to those who are not their employees under section 4(2).[75]

One of the problems in the setting of standards, such as the standard concerning premises, plant and substances in the present section, is the question of standards for vulnerable or disabled groups, *e.g.* immigrants who cannot read English or disabled persons for whom special precautions, *e.g.* ramps, may be needed. It would be odd if employers, who are required to accept (at least in theory if not in practice) a quota of registered disabled persons, could set standards by reference to non-disabled persons. On the other hand the "measures" which might need to be taken "by persons using the premises" or the plant or substances provided therein must be "reasonable" for a person in the position of the "controller" and must be "reasonably practicable." It seems to be undeniable that cost must play a major part in such circumstances. Access to and facilities at public, university and school buildings are governed by sections 4 and 8 of the Chronically Sick and Disabled Persons Act 1970 to which must be added requirements as to access to and within premises, parking facilities and sanitary conveniences (if any) where the premises are covered by the Offices Shops and Railway Premises Act 1963, or are factories as defined in section 175 of the Factories Act 1961, provided in each case that persons are employed on work therein.[76]

(e) *The working environment: Section 2(2)(e)*

The requirement in section 2(2)(*e*) that the employer must provide and maintain "a working environment" for his employees that is, so far as is reasonably practicable, safe, without risks to health, and that he furnishes adequate facilities and arrangements for their welfare at work, is new. It recognises the entitlement of all workpeople to reasonable minimum standards of *health and amenity in their working environment* (Robens Report, para. 170, *italics supplied*). In one sense the provision is not new, except to generalise several health and welfare provisions in the existing codes, such as Part I (Health), Part II (Safety) and Part III (Welfare) of the Factories Act 1961 although, once again, the provision will be "new" to the "new entrants." Failure to provide a supply of drinking water or accommodation for clothing does not raise health or safety issues but does concern "amenity," to use the Robens term.

[75] See *Aitchison* v. *Howard Doris Ltd.* [1979] S.L.T. 22.
[76] Added by s.2, Chronically Sick and Disabled Persons (Amendment) Act 1976 and amended by s Disabled Persons Act 1981.

WRITTEN POLICY STATEMENT: SECTION 2(3)

The "self-regulation" recommended by the Robens Committee in preference to "negative regulation by external agencies" must be made to happen; it cannot be guaranteed to occur as a sort of commanded spontaneous combustion. Like the prospect of being hanged on a Monday morning, the intention of section 2(3) is to concentrate the mind. Except in prescribed cases, it is the duty of every employer to prepare (and keep revised) a written statement of his general policy with respect to the health and safety at work of his employees and the organisation and arrangements for the time being in force for carrying out that policy, and to bring the statement and any revision to the notice of all of his employees.

In line with the current policy of giving incentives and dispensations to small firms, the Employers' Health and Safety Policy Statements (Exception) Regulations 1975 (S.I. No. 1584) exempt any employer who carries on an undertaking in which for the time being he employs less than five employees from the requirements relating to the written policy statement. The phrase "the time being" means at any one time so that, for example, a companay employing twelve people on a three-shift system under which only four work at any one time will be exempt from the statutory duty with regard to the policy statement. Similarly, an employer who employs five employees, of whom two are relief staff (so that less than five are employed at any one time), would be covered by the dispensation.[77] The rationale of the dispensation is that communication in small firms tends to be oral rather than written.

The inspectorate devoted much time and effort, particularly in the early days of the Act, in assisting employers, particularly new entrants such as universities and hospitals, to prepare written statements; this exercise involved measures to deal with identified risk and, arguably more important, the organisational responsibility for health and safety within the enterprise. The HSC issued a new leaflet on the new duty but wisely refrained from circulating a model "blue-print" for health and safety which might have been copied without any thorough assessment of possible hazards. Although it is true that most of the improvement and prohibition notices issued under the Act relate to the so-called "physical factors," it was open to the enforcing authorities to use their powers, including the power to issue notices, for the purposes of ensuring compliance with

[77] *Osborne* v. *Taylor (Bill) of Huyton Ltd. The Times,* December 8, 1981, D.C. See also *Osborne* v. *Bill Taylor of Huyton* [1982] 1 R.L.R. 17, D.C.

requirements as to written policy statements (or information, instruction, training and supervision under s.2(2)(*c*), above). Because these were new requirements, in contrast to the old practice of posting up abstracts, the enforcing authorities have preferred advice and persuasion to the big stick of enforcement notices and, *a fortiori* criminal prosecution.

During its parliamentary passage, the Act received a new provision (now s.79) which amends the Companies Acts to empower the Secretary of State to make regulations concerning prescribed classes of companies requiring the inclusion in directors' reports of such information as may be prescribed about the arrangements for securing the health, safety and welfare at work of employees of the company and its subsidiaries, and for protecting other persons against risks to health or safety arising out of or in connection with the activities at work of those employees. As it was put in Parliament, it was thought that the new provision (one recommended by the Robens Committee) "would concentrate the minds of chairmen and directors most wonderfully if the company's health and safety performance had to be accounted for in public."[78]

SAFETY REPRESENTATIVES AND SAFETY COMMITTEES: SECTION 2(4)–(7)

One of the most radical innovations in the Act so far as those outside mining were concerned was the provision (in section 2(4)–(7)) for the appointment of safety representatives and safety committees who are intended to play a central role in the new régime of "self-regulation." Because of their importance they are discussed in detail (see Chapter 8, 190). In its original incarnation, section 2(5) provided for the election of safety representatives by employees in those cases in which there was no trade union or recognised trade union; but this statutory franchise was taken away by the repeal of that subsection by the Employment Protection Act 1975, a repeal which resulted in criticisms of the Government because of pro-union bias and indifference to the plight of non-unionised workers. The Government's reply was that it was open to employers and non-unionised workers to make their own arrangements if they so wished, although it is clear that this right to agree on safety representatives is different from the "unilateral right" conferred by section 2(4) (which relates only to appointments by recognised trade unions).

[78] H.C. deb., June 18, 1974, col. 241. No Regulations have been made at the time of writing.

DUTIES OF EMPLOYERS AND SELF EMPLOYED TO PERSONS OTHER THAN THEIR EMPLOYEES: SECTION 3

The introduction of this section marked a new departure in occupational health and safety regulation, both in this country and throughout the world. The seeming absence of concern on the part of British health and safety inspectorates for *public* wellbeing had given rise to sustained and authoritative criticism.[79]

The Robens Committee considered that the inspectorates believed that if employees were afforded the correct degree of protection the protection of the public would then automatically follow. The Committee did not share this belief. Sometimes, there were no employees, as in coin-operated launderettes, for which there should be separate provision for public safety. In addition, the general public could be placed at risk from external sources, such as the use of an explosive substance in quarrying and from internal dangers, such as the malfunctioning of a lift or a dangerous machine in use in a department store.

Moreover, the Committee were much concerned with the possibility of a large-scale disaster which might arise from the storage of petroleum, liquified petroleum gas, liquid oxygen, chlorine, phosgene and sulphur dioxide. Unfortunately, within two years of the Committee submitting their Report, this concern was shown not to have been misplaced in that just such a disaster was to occur at Flixborough, near Scunthorpe, where 28 people were killed and one hundred and five injured. It was Britain's biggest explosion since the war; its consequences were largely unforeseen and the cost of the damages was an estimated £40 million.[80] Since then there has been a continuing debate both in Parliament and generally concerning the dangers and risks to health presented by various installations, both existing and proposed on Canvey Island and the neighbouring part of Thurrock.[81]

A disaster with lasting consequences was to occur in Seveso, Italy, in 1976 when a runaway reaction during the production of triphenol resulted in the discharge of products containing TCDD[82] over an area of 2.8 Km^2 (700 acres). Four per cent. of

[79] Report of the Investigation of the Crane Accident at Brent Cross Hendon, June 20, 1964, Cmnd. 2768, H.M.S.O., 1965; Report of the Tribunal appointed to inquire into the Disaster at Aberfan, October 21, 1966, HL 316, HC 553, H.M.S.O., 1967. For criticism of the enforcement policy of H.M's Inspectors of Alkali and Clean Air in relation to lead pollution at Avonmouth, see *The Listener*, May 2, 1974, pp. 559–562.

[80] See The Flixborough Disaster—Report of the Court of Inquiry H.M.S.O. 1975 and Chapter on Major Hazards page 228.

[81] See Canvey: an investigation of potential hazards from operations in the Canvey Island/Thurrock area 1978 and Canvey: a second report. A review of potential hazards from operations in the Canvey Island/Thurrock area three years after publication of the Canvey Report, 1981.

[82] 2, 3, 7, 8 tetrachlorodibenzodioxin.

the domestic animals living in the contaminated zones died spontaneously and the remaining 77,716 animals were slaughtered as a preventive measure to protect the food chain. Some 736 people were exposed to relatively high doses of TCDD and it is probably too early to estimate the damage caused except to say that an unusually high incidence of chloracne was detected in the area of the contaminated townships.[83]

Combined with the fear of a larger scale accident described above was the need to provide for an employer's duty towards his subcontractors and for workpeople employed by those subcontractors and for hazards to the public on the highway from scaffolding collapse, for example, and for hazards to visitors, both able-bodied and infirm, adults and children authorized and unauthorized, who find themselves in a workplace.

Section 3 provides that employers and the self employed[84] should conduct their undertakings in such a way as to ensure that those who are not in their employment should not be exposed to risks to their health and safety.

This duty on an employer to conduct his undertaking in such a way as to ensure that persons not in his employment who may be affected thereby are not exposed to risks to their health and safety is wide enough to include the duty to provide information and instruction.[85] In addition, in order to satisfy the general duty imposed by section 2(1), Health and Safety at Work etc. Act 1974, an employer is required to provide a safe system of work for, and to inform and instruct, employees of other persons not in any contractual relationship to him where that requirement is necessary to ensure the health and safety of his own employees.[86]

In *Aitchison* v. *Howard Doris Ltd.*,[87] a main contractor was held not to be liable to the employee of a subcontractor under section 3(1) who had suffered an injury when his leg was trapped between a cement barge and a boat. The High Court of Justiciary held that section 3 did not cover safety of access which is under the direct control of a third party. The Court was clearly of the opinion that if any section applied it was section 4 (below). It would seem that the requirement as to the conduct of an undertaking has a restrictive effect, so that if a person is to be

[83] See further The Seveso Accident: its nature, extent and consequences *Ann. occup. Hyg.* Vol. 22 pp. 327–370, 1979.

[84] The term "self employed person" is modified by the Health and Safety (Genetic Manipulation) Regulations 1978, Regulation 4, in relation to any activity involving genetic manipulation, so that 'self employed person' includes a reference to any person who is not an employer or employed person in relation to that activity.

[85] *R.* v. *Swan Hunter Shipbuilders Ltd.* [1981] Crim L.R. 833.

[86] *R.* v. *Swan Hunter Shipbuilders Ltd. ibid.* See the portmanteau duty above.

[87] [1979] S.L.T. (Notes) 22.

made liable for premises not used in the conduct of the undertaking it must be under section 4, which uses the test of control.

Section 3(2) provides that it shall be the duty of every self employed person to conduct his undertaking in such a way as to ensure, so far as is reasonably practicable, that he and other persons (not being his employees) who may be affected thereby are not thereby exposed to risks to their health and safety.

Section 3(3) provides that in such cases as may be prescribed,[88] it shall be the duty of every employer and every self employed person, in the prescribed circumstances and in the prescribed manner to give to persons (not being his employees) who may be affected by the way in which he conducts his undertaking the prescribed information about such aspects of the way in which he conducts his undertaking as might affect their health or safety. In the Swan Hunter Case, the failure of Swan Hunter to give Telemeter employees information and instruction concerning the safe use of oxygen was also a breach of section 3(1) of the Act. The duty under section 3(3) (above) does not limit the ambit of section 3(1). Section 3(3) deals with a limited range of prescribed cases in which prescribed information has to be given and does not impinge upon section 3(1) which is wide enough to include the giving of information and instruction to employees other than one's own employees.

Section 3(1) as a whole is thought to have been made to obviate the occurrence of large scale hazards and neighbourhood risks (see the discussion of Flixborough, above) in which connection the proposed Hazardous Installations (Notification and Survey) Regulations will be of great assistance (see below, Chapter 11, p. 238). The subsection is wide enough to include controls on isolated occurrences, such as the use of laser beams to provide the Christmas illuminations in Oxford Street, London.[89]

DUTIES OF CONTROLLERS OF PREMISES, ETC.: SECTION 4

Section 4 shows that the Act has not been able to escape entirely from the "premises" test and the "activities" test which the Robens Committee found to be so restrictive and technical in their operation. The "premises" test depends upon where a person is doing something and not upon what he is doing. The "activities" test necessarily involves lists of activities and processes in need of constant revision. The Act in the main is

[88] No regulations have yet been made under this provision.
[89] Health and Safety at Work (1979), vol. 1, p. 70.

based upon the employment relationship but contains provisions which are a recognition of the need to impose duties on persons other than employers. Section 4 breaks away from the employment relationship by imposing duties on those who control premises (their access and egress included) or plant or substances in such premises to take reasonable measures to ensure so far as is reasonably practicable that the premises, plant or substance is, or are, safe and without risk to health. The elastic concept of "reasonableness" applies twice in this section: first, in requiring the "controller" to "take such measures as it is *reasonable* for a person in his position to take" to ensure, secondly, so far as is *reasonably* practicable that the premises and so forth are safe and without risks to health.

The duties are based upon the control of premises (plant or substance) and such control may be "to any extent." If a person has obligations by virtue of any contract or tenancy either to maintain or repair premises covered by the section or in connection with the safety of or absence of risk to health arising from plant or substances in any such premises, he is a "controller" for present purposes. The use of control in preference to ownership or occupation casts the net wider, although it should be noted that the overall requirement is that the control must be by way of trade or business or other undertaking (whether for profit or not) so that, for example, a householder would not be liable as such under the section. The word premises normally connotes buildings or land, but in the present context includes any place and in particular any vehicle, vessel, aircraft or hovercraft as well as land, offshore and other installations.

An example of the application of section 4 is provided by *Northampton B.C.* v. *Farthingstone Silos Ltd.*,[90] in which a local authority successfully prosecuted the controllers of premises when an electrician fell sixteen metres to his death whilst carrying out repairs to a motor at the top of a grain dryer. The health and safety inspector found that the metal treads to the stairway were bent, that the handrails were unsafe and that the floor, upon which the electrician had to stand and work, some 45 feet above ground had some footplates which were missing and some which were loose. The accident occurred when the electrician stepped back on to the unsupported end of a steel floor-plate from which he fell to the ground. It was held that the controllers of the site were in serious breach of section 4 for which they were fined £2,000 plus costs. In the Aitchison Case (discussed above[91]), the Court considered that the main contrac-

[90] Northampton Crown Court (1981), unreported.
[91] See p. 86, above.

tor might have been liable under section 4 if it had been proved that he was in control of the vessels between which the access existed; but since neither the section nor control had been libelled in that case the complaint was held to be irrelevant.

The duties are owed to persons who are not employees of the "controller" but who use non-domestic premises made available to them as a place of work or as a place where they may use plant or substances provided for their use there. Non-domestic premises are defined by exclusion as premises other than those occupied as a private dwelling (including any garden, yard, garage, outhouse or other appurtenance of such premises which is not used in common by the occupants of more than one such dwelling).[92] The exclusion of domestic premises, including domestic service, from the Act accords with the Robens view that the law should not encroach upon private individuals within their own homes. On the other hand, the "domestic system" is not dead in that there is currently considerable concern at the widespread use of "homeworkers" whose safety and health may be at risk. If the homeworker is an employee, the employer may well be affected by section 2 which has broken free from the geographical constraints in previous legislation. In this context, it may be noted that the definition of "employee" in s.53(1) is narrow in that, unlike certain statutes, it does not refer to those who contract "personally to execute any work or labour."[93] If the homeworker is not an employee it is difficult to see how the Act will apply. The HSC is aware of the problem and has issued a consultative document and draft regulations to cover homeworkers who the Robens Committee considered should be covered by requirements as to safe materials and equipment together with adequate instruction. The Government has been advised that the Act covers students and trainees at institutions of further education, including Government training establishments.[94] The position of students is curious because in relation to their educational institution they are neither employees nor members of the public. They may be affected by the conduct of an undertaking under section 3 or by those who control premises, plant or substances under section 4, although it is interesting to note that students are not covered in relation to the use of non-domestic premises made available to them as a "place of work" under section 4(1)(*b*) in that a laboratory, for example, might be their place of work in the ordinary sense but

[92] s. 53(1).
[93] See the definition of "workman" in s.8, Industrial Courts Act 1919: of "worker" in s.28, Wages Councils Act 1979 (*cf. Westall Richardson Ltd.* v. *Roulson* [1954] 1 W.L.R. 905).
[94] Written Answers, Parl. Deb. H.C., 2 August 1976, co. 550.

is not their place of work within the meaning given to that term in the Act.[95]

DUTY OF PERSONS IN CONTROL OF CERTAIN PREMISES IN RELATION TO HARMFUL EMISSIONS INTO THE ATMOSPHERE: SECTION 5

Although the terms of reference of the Robens Committee specifically excluded considerations of the question of general environmental pollution, its members were firmly of the view that common control arrangements should apply and that it made little sense to have two inspectors visit a workplace with one to deal with emissions within the workplace affecting its workpeople and the other to deal with emissions affecting the health of the general public.

Section 1(1)(*d*) provides that the provisions of Part I of the Act shall have effect with a view to controlling the emission into the atmosphere of noxious or offensive substances from premises of any class prescribed for the purposes of this paragraph.

Whilst section 5(1) provides that it shall be the duty of the person having control of any premises of a class prescribed[96] for the purposes of section 1(1)(*d*) to use the best practicable means for preventing the emission into the atmosphere from the premises of noxious or offensive substances and for rendering harmless and inoffensive such substances as may be so emitted.

In section 5(1) the reference to means includes a reference to the *manner* in which the plant provided for those purposes is used and to the *supervision* of any operation involving the emission of the substances to which that subsection applies.[97]

Regulations prescribing a substance as noxious or offensive are conclusive on that point.[98]

Any reference in section 5 to a person having control of any premises is a reference to a person having control of the premises (not employees therefore) in connection with the carrying on by him of a trade business or other undertaking (whether for profit or not) and any duty imposed on any such person by this section shall extend only to matters within his control.

The duty in section 5 is to use the best practicable means, and although it is not defined in the Health and Safety at Work etc.

[95] See the meaning of "work" in s.53(1), which may be extended by Regulations (s.53(2)).
[96] No regulations have yet been made under this provision.
[97] Section 5(2), echoing section 27(1) Alkali etc. Works Regulation Act 1906.
[98] Section 5(3) Noxious and offensive substances includes liquids and gases. See section 53. The section is aimed primarily at the control of toxic dusts, but it goes beyond matters which are harmful to health. It would be possible to make regulations dealing with nuisances under this provision.

Act 1974 it is defined in section 72 of the Control of Pollution Act 1974 for the purposes of Part III (Noise).

Section 72(2) provides that "practicable" means reasonably practicable having regard among other things to local conditions and circumstances, to the current stage of technical knowledge and to the financial implications.

Section 72(4) further provides that the test of best practicable means is to apply only so far as compatible with any duty imposed by law, and in particular is to apply to statutory undertakers only so far as compatible with the duties imposed on them in their capacity of statutory undertakers.

Interestingly section 72(5) provides the link with the provisions of the Health and Safety at Work etc. Act 1974 in that the test of best practicable means is to apply only so far as compatible with safety and safe working conditions, and with the exigencies of any emergency or unforeseeable circumstances.

Additionally section 27(1) Alkali etc. Works Regulation Act, 1906 provides that "best practicable means" refers not only to the provision and efficient maintenance of appliances adequate for preventing the escape but also to the manner in which they are used and to the proper supervision, by the owner, of any operation in which the gases etc. are evolved.

Best practicable means is a higher standard to achieve than so far as is reasonably practicable. In *Scholefield* v. *Schunck,*[99] a case decided under Factories Act 1844, it was held that it was not enough that the precautions ordinarily adopted in the trade had been observed. What was required was the best available means which could be adopted for securing the end in view.[1]

A useful policy note on the inspectorate's view of the meaning of best practicable means is contained in the 103rd Annual Report on Alkali, etc., Works by the Chief Inspectors 1966.

Alkali and Clean Air Inspectors have taken the view, over the years, that co-operation is better than coercion and confrontation in seeking improvements in the modernisation of plant and equipment. In some cases, for example where plant and equipment is new, a high standard will be demanded but, in others, where for example no practicably efficient means for prevention have been found or where there have been no justified complaints and works have equipment installed which was in place before the Act came into effect, they will be allowed to operate that plant and equipment until the end of its economic life.

[99] (1855) 19 J.P. 84.

[1] See also *Manchester Corporation* v. *Farnworth* [1930] A.C. 171.

Full "best practicable means" is reached "when the standard of treatment of emissions is so high as to result in little or no impact on the community and with no scope for further major improvements."

GENERAL DUTIES OF MANUFACTURERS, ETC.: SECTION 6

There is little doubt that prior to 1974, protective legislation, with certain exceptions,[2] tended to concentrate on the person in charge of the enterprise, such as the factory occupier, who more usually than not would be an employer. In this respect, the statutes tended to follow the common law which had worked out a comprehensive régime of liability for employers towards their employees but which was more reluctant to saddle manufacturers and suppliers with liability. The common law was, of course, concerned with loss, liability for which was (and is) determined on the basis of causation, whereas the whole purpose of preventive legislation is to deter rather than to compensate. When the House of Lords in *Davie* v. *New Merton Board Mills Ltd.*[3] reduced the liability of employers at common law with regard to equipment obtained from a manufacturer or supplier, the response of the Legislature was to make the employer liable for harm or loss caused by defective equipment, provided that negligence could be shown, for example, on the part of the manufacturer—a sort of transferred negligence."[4] Of course, an employer who kept in use a piece of equipment known to be dangerous could be held liable for negligence because this might break the "chain of causation" back to the negligent manufacturer or supplier, as in *Taylor* v. *Rover Co. Ltd.*[5] where the makers of the tool (who had made it to the employer's specification) were entitled to assume that a third party (responsible for hardening the metal) had done his job properly. On the civil side, we have imported from the U.S.A. the new subject of "product liability," now reinforced by an EEC Directive on that subject, with parallel statutes having preventive objectives and backed up by criminal sanctions, such as the Consumer Safety Act 1978. It is true that there are isolated provisions which move away from the workplace to the manufacturer or supplier, such as section 17 of the Factories Act 1961, which imposes limited duties with regard to the construction,

[2] Manufacture of tractors with safety cabs: approval of Secretary of State needed for certain mining equipment.
[3] [1959] A.C. 604.
[4] Employer Liability (Defective Equipment) Act 1969.
[5] [1966] 1 W.L.R. 1491.

sale or letting on hire of machines in factories intended to be driven by mechanical power.[6]

One theoretical solution to the supply of dangerous articles and substances would be some state licensing agency but, with certain exceptions, this is not a practicable idea. Something as innocuous as a chair may collapse but it would be difficult to justify a Chair Licensing Authority. Where planning approval is needed for a major hazard, consultation of HSE involves a degree of "pre-natal control," and this would be considerably extended if the second report of the Advisory Committee on Major Hazards were to be adopted. This report gives details of a possible licensing system for plant of the greatest hazard and explains the thinking behind proposed notification and hazard survey regulations.[7]

Section 6 deals with articles and substances "for use at work," whereas the Consumer Safety Act 1978, according to its preamble, deals with "the safety of consumers and others." There is a requirement that in the making of safety regulations under this latter Act concerning goods suitable for use at work there must be consultation with the HSC (see section 1(4)). Clearly, if "others" includes those who use things at work, there is room here for some overlap and it is this consideration which led the Secretary of State for Employment to make clear to HSC in a letter (dated 7 October 1975) that consumer safety lay outside the HSC remit. There is no definition of "consumer goods" in the 1978 Act and some goods, *e.g.* portable electric drills, might be used by consumers *and* by workers. In this latter use there might also be liability under sections 2 and 4 of the 1974 Act.

The overall purpose of section 6 is to set out an extended "chain of responsibility" involving, not only the employer, but manufacturers, suppliers, erectors and installers. The new duties are complementary to those set out in prior legislation, or contained in other provisions of the 1974 Act, *e.g.* the duties under section 2. These new duties extend only to things done in the course of a trade, business or other undertaking carried on by a person (whether for profit or not) and to matters within that person's control.[8] So that, for example, a manufacturer who took measures which were reasonably practicable, such as attaching warnings to his product, would not be responsible for some "intermediary" who stripped off the warning and disposed of it

[6] *Biddle* v. *Truvox Engineering Co. Ltd.* [1952] 1 K.B. 101, in which civil redress was refused. Prosecutions under section 17 Factories Act 1961 were relatively few. See *The Use of Standards in Support of Health and Safety Legislation* (1980) J.H. Locke.

[7] See Chapter 11 The Control of Major Hazards (p. 228).

[8] s.6(7).

before supplying the product to a user for use at work. The "chain of responsibility" might not end with the user since the latter might dispose of waste products, by-products or worn-out machinery (a not uncommon spectacle in the so-called "de-industrialisation of Britain") and, if these are supplied for use at work, liability may well be incurred.

Section 6 deals with "articles" and "substances" for use at work, the subsections on the latter *mutatis mutandis* largely duplicating the former.

"Articles"

An "article for use at work" means any plant designed for use or operation (whether exclusively or not) by persons at work, and any article designed for use as a component in any such plant. The term plant includes any machinery, equipment or appliance.[9] Under section 6(1) it is the duty of any person who designs, manufactures, imports or supplies any article for use at work—

(*a*) to ensure, so far as is reasonably practicable that the article is so designed and constructed as to be safe and without risks to health when properly used;

(*b*) to carry out or arrange for the carrying out of such testing and examination as may be necessary for the performance of the duty imposed upon him by (*a*) above;

(*c*) to take such steps as are necessary to ensure that there will be available in connection with the use of the article at work adequate information about the use for which it is designed and has been tested, and about any conditions necessary to ensure that, when put to that use, it will be safe and without risks to health.

The section only applies to articles (and substances) for use at work which means that articles and substances which are stock-in-trade are not included. Thus, where reconditioned engines are purchased by a garage for installation in the vehicles of customers and these disintegrate whilst under test, such engines would not be articles for use at work but would be the stock-in-trade of the garage. In the factories legislation the duty to fence dangerous parts of machinery has been held not to extend to machines or parts of machines which are themselves products of the factory.

It is curious that the requirement in (*a*), above, as to design and construction applies, not only to the designer and manufac-

[9] s.53(1), but not buildings—architects escape!

turer, but also to the importer or supplier, whose knowledge of design and construction might be little or non-existent. Presumably, it would be reasonably practicable in many cases for the latter to make enquiries of the manufacturer as to the safety and health standards in the design and construction of the article. Clearly, the designer and manufacturer is better able to "ensure" compliance with the required standards than importers or suppliers.[10]

Similarly, the testing and examination of articles under (*b*), above, is a responsibility lying upon all four categories (designer, manufacturer, importer and supplier), although, in practice, most testing and examination will take place either at the design stage (*e.g.* by using a prototype) or, more usually, at the manufacturing stage. However, section 6(6) (which applies also to substances) renders it unnecessary for a person (*e.g.* an importer, supplier, erector or installer) to repeat testing, examination or research which has already been carried out (*e.g.* by a manufacturer), provided it is reasonable for him to rely on the results thereof. If, however, an importer were to deliver dangerous machinery without first examining it he would not, on those facts alone, have a defence under the subsection. It would not be possible for a designer or manufacturer to say that he could not carry out testing or research because of the expense, just as it would not lie in the mouth of an importer or supplier to "pass the buck" by saying that he relied on information given by a manufacturer when it was not reasonable to rely on that information. In particular where articles or substances are known to be inherently dangerous, information should be available about the hazards and about the precautions which should be taken in storage and handling.

Designers, manufacturers, importers and suppliers are obliged under (*c*) above to secure that adequate information will be available to the user about an article, the use for which it was designed and tested and any conditions necessary to ensure that, when put to that use, it will be safe and without risks to health. The information may be given in the form of an instruction manual or data sheet, or refer the user to a standard in a Code or other authoritative source including a technical standard such as those in British Standards. If the required information is made available to the user, the latter is able to discharge his own duty to his employees under section 2(2)(*c*) which deals with the provision of information, instruction, training and supervi-

[10] Where there is a choice in relation to enforcement action the Inspector should ask: which of several people has the matter most under his control? Research and development firms might be included if the error which caused the product to be unsafe originated with them.

sion (see p. 78, above). Information will probably be available only if the user's attention has been drawn to it either in the sale documents or in the service manual and if he has been given a name and address to approach for further information. Section 6(10) protects a designer, manufacturer, importer or supplier who has provided adequate information.

Section 6(8) furnishes a *locus poenitentiae* to designers, manufacturers, importers or suppliers of articles[11] for use at work, any of whom has supplied an article for or to another on the basis of a written undertaking by that other to take specified steps sufficient to ensure, so far as is reasonably practicable, that the article will be safe and without risks to health when properly used. Such an undertaking will serve to relieve the former from the duty in (*a*), above, to such extent as is reasonable having regard to the terms of the undertaking—flexibility is permissible. It will be noted that the written undertaking does not relieve a person from his duty under (*c*), although the relief with respect to duty (*a*) to some extent confers relief with respect to duty (*b*) in so far as (*a*) and (*b*) are interrelated. The intention is not to allow a person to "pass the buck," but to deal with cases in which it is reasonable to relieve a person of his liability under (*a*) above, *e.g.* where he constructs a machine to a buyer's specification on receiving a written undertaking from the latter that it will be securely fenced, provided of course that it was reasonable to act on that undertaking. By virtue of section 101 of the Magistrates' Court Act 1980, where a defence is based on an exception, exemption, proviso, excuse or qualification the burden of proving the latter falls upon the defendant.

Those who undertake the design or manufacture of any article for use at work are obliged by section 6(2) to carry out or arrange for the carrying out of any necessary research with a view to the discovery and, so far as is reasonably practicable, the elimination or minimisation of any risks to health or safety to which the design or article may give rise. The HSC/HSE devote a considerable proportion of their budget to research, the results of which, together with the reports of Joint Standing and Advisory Committees, are available to designers and manufacturers upon whom, however, the subsection places the duty relating to research.[12]

Section 6(3) extends the chain of responsibility to include those who erect or install any article for use at work in any

[11] Note that substances are not covered.

[12] *Cf.* s.11(2)(*a*) for the duty of the HSC to make such arrangements as it considers appropriate for the carrying out of research and the publication of the results of such research.

premises where that article is to be used by persons at work. They must ensure so far as is reasonably practicable that nothing about the way in which the article is erected or installed makes it unsafe or a risk to health when properly used. The subsection is based on acceptance of the possibility that an article may be well designed and carefully constructed yet constitute a risk to safety or health because it has been badly erected or installed. Would an erector or installer incur criminal liability if he erected or installed the article carefully, but left it without some ancilliary safety device such as a guard or exhaust extractor? The HSC considers[13] that the duty extends to these ancilliary matters and whilst this might well be so, it is possible that the contract with the erector or installer expressly excludes additional devices on the basis that separate arrangements will be made with regard to these. The erector or installer would be free from criminal liability if the article were to be used improperly and for this purpose an article (and the same applies to a substance) is not to be regarded as properly used where it is used without regard to any relevant information or advice relating to its use which has been made available by a person by whom it was designed, manufactured, imported or supplied.[14]

"Substances"

Section 6(4) and (5) specify the duties of manufacturers, importers, or suppliers of any substance for use at work. The former subsection repeats, *mutatis mutandis*, the three duties (*a*), (*b*) and (*c*), set out above in connection with articles and to which the reader is referred (p. 94 above). A "substance" is defined by section 53(1) as "any natural or artificial substance, whether in solid or liquid form or in the form of a gas or vapour," a definition which breaches the schoolmaster's injunction against circularity. Subsection (5) repeats the duty, again *mutatis mutandis,* with regard to research in relation to articles. Although the duties are cast in practically identical terms, the safety and health risks arising from substances differ from those arising from articles in that they raise issues of toxicity, disease, chemical reaction and the like. A vast amount of work has been done by HSC/HSE and specialist committees on these risks and much valuable guidance has been made available on safe limits, such as Threshold Limit Values (TLV's) for certain substances. The duty on manufacturers of substances for use at work to carry out or arrange for the carrying out of any necessary

[13] Guidance Note Gs 8.
[14] *Ibid.*

research with a view to the discovery and so far as is reasonably practicable the elimination or minimisation of any risks to safety or health to which the substance may give rise extends to carcinogens.[15]

One cannot use the case of *Vacwell Engineering Ltd.* v. *B.D.H. Chemicals Ltd.*[16] as a precedent *sub silentio* on points not discussed in that case, but it would seem that the approach in that case would be followed in relation to section 6. Vacwell were manufacturers of plant and equipment designed to produce transistor devices and called upon B.D.H. to supply chemicals needed for that work. In 1964, B.D.H. wished to introduce a new chemical, boron tribromide, to the market and advertised it in their catalogue. Both Vacwell and B.D.H. knew that boron tribromide reacted with water, but neither knew that the reaction was violent. Vacwell ordered 400 glass ampoules containing the chemical from B.D.H. in 1966 and these were supplied with labels bearing the words "harmful vapour." A few months later, when two physicists were engaged in washing these ampoules prior to using them in the manufacturing apparatus, an explosion occurred killing one of the two men and causing extensive damages to Vacwell's premises. The likelihood is that the ampoules were accidentally broken whilst being washed. It was held, *inter alia*, that B.D.H. had failed to provide and maintain a system for carrying out adequate research into scientific literature to ascertain known hazards and they had additionally failed to carry out adequate research into the literature available to them concerning boron tribromide (in looking at recent literature they had overlooked earlier reports on the explosive potentiality in relation to water).

The generality of section 6(4)(*c*), which lays down a sweeping requirement as to making available adequate information concerning substances and their use at work, contrasts with (but is not replaced by) the particularity in certain regulations, such as those which deal with the packaging and labelling of dangerous substances and which implement an EEC Directive.[17] These regulations impose duties with regard to the supply of some 800 prescribed dangerous substances (and there are more to be prescribed in the future) which must be packaged and labelled in accordance with the regulations, which utilise graphic illustrations and easily-understood symbols. According to HSE Guidance Notes, "supply" means "to pass to another person or organisation by sale, lease, hire, hire-purchase or as a

[15] Parl. Deb. H.C., Written Answers, November 17, 1976, col. 597.
[16] [1971] 1 Q.B. 111.
[17] Packaging and Labelling of Dangerous Substances Regulations 1978 (S.I. 1978 No. 209), as amended by S.I. 1981 No. 792.

commercial sample," but would not include transfer solely for warehousing or as part of distribution. Enforcement lies mostly with the HSE, although registered pharmacies and shops come under the Pharmaceutical Society of Great Britain and local weights and measures authorities, respectively. Consumer safety is the chief object of regulations made under section 1 of the Consumer Safety Act 1978, whilst public safety is the concern of the Dangerous Substances (Conveyance by Road in Road Tankers and Tank Containers) Regulations 1981.[18]

Suppliers of Articles and Substances

Articles or substances may be supplied by way of sale, lease, hire or hire-purchase, whether as principal or agent.[19] The essence of supply is a commercial transaction as opposed to the supply of something by way of gift or as a friendly loan. Section 6(9), however, draws a distinction between an "ostensible supplier" who supplies an article or substance under a hire-purchase agreement, conditional sale agreement or credit-sale agreement, as a financing device in order to permit a third person to acquire the article or substance from an "effective supplier." In a typical hire-purchase arrangement, the effective supplier transfers ownership to the ostensible supplier (the finance company) so that the latter can permit the third person to acquire the article or substance by paying the instalments and option to purchase. For the purposes of section 6, the duties are cast, not upon the ostensible supplier, but upon the effective supplier. The "exoneration" of the ostensible supplier was extended by regulations to those who supply an article for use at work to a customer under a lease, an arrangement which has become increasingly popular in recent years.[20]

Enforcement of Section 6

Inspectors have powers under section 20(2)(*h*) to cause articles or substances found on any premises which they have power to enter and which they think cause danger to health or safety to be dismantled or subjected to any process or test, although they cannot damage or destroy them unless to do so is necessary for the purpose of enabling them to carry into effect any of the relevant statutory provisions. The enforcement of section 6 involves visits by inspectors to manufacturers and importers

[18] S.I. 1981 No. 1059.
[19] s.53(1).
[20] Health and Safety (Leasing Arrangements) Regulations 1980 (S.I. 1980 No. 907).

who are advised of the law and of the standards required to comply with the law. Some 25 per cent. of known manufacturers and some 50 per cent. of known importers are visited each year. Visits are also paid to users, particularly where defects in use have appeared. As the Director General of HSE has said, "most problems and accidents arise from the use made of a product rather than from defects in the product itself" so that "only a small proportion of the total number of enforcement initiatives are against manufacturers or importers."[21]

The table below shows the numbers of enforcement notices associated with section 6.[22]

	Improvement Notices	Prohibition Notices
1976	7	4
1977	17	16
1978	17	10
1979	32	7

The distribution of prosecutions under section 6 as between designers, manufacturers, importers and suppliers or between the foregoing and users does not appear to have been published, although the prosecutions under the section by HSE inspectors have been published. The details, given below, do not record local authority prosecutions and refer to informations laid (one case may include several informations).[23]

	Informations			
	Laid	Convictions	Withdrawn	Dismissed
1975	3	3	—	—
1976	11	11	—	—
1977	28	23	—	5
1978	32	23	3	6
1979	28	22	5	1
1980 (Jan. to Nov.)	25	22	—	3

The importance of section 6 is reflected in the increasing amount of time which HSE is devoting to seeing that articles and substances for use at work are safe, whilst dealing "evenhandedly between domestic manufacturers and suppliers of imported equipment."[24]

21 *The Use of Standards in Support of Health and Safety*, Mr. J.H. Locke, op. cit.
22 Parl. Deb. H.C., Written Answers, February 16, 1982, col. 99.
23 Based on HMFI and HMAI records: Parl. Deb. H.C., Written Answers, 17 March 1981, col. 69.
24 Director General's Report to HSC (1980–81), para. 2.

GENERAL DUTIES OF EMPLOYEES AT WORK: SECTION 7

Because of his control of the enterprise, it is natural and logical that the main thrust of health and safety law should bear upon the employer. An employer may install a safe system of work with every conceivable health and safety device and yet have his carefully-laid plans frustrated by the foolishness or indifference of his workforce. So far as the compensatory role of the common law duty of care is concerned, a distinction has been drawn between the "personal" duty of the employer (safe system, safe appliances and proper personnel) which is non-delegable and the "vicarious" liability of the employer for personal injury caused by an employee in the course of employment. In practice, an employee is much more likely to be injured by the act or omission of a fellow-employee than by an employer's breach of his "personal" duty, although it was some time before the common law accepted the employer's vicarious liability to an employee injured by the fault of a fellow-employee with whom he was in common employment. The common law recognises the reality of fault on the part of employees in various ways: by reducing compensation because of the contributory negligence of the employee, by defeating a claim for compensation because the employee has consented to the risk (in which connection a breach of statutory duty by the employee may well be highly relevant) and, so far as the contract of employment is concerned, by the implication of a term that the employee must both have the skill which he warrants and exercise due care. A good example of the way in which employees deliberately run risks is to be found in a leading civil case in which the issue of consent to the risk (*volenti non fit injuria*) was discussed including a breach of statutory duty by the two employees concerned—the case of *Imperial Chemical Industries Ltd.* v. *Shatwell* where Lord Pearce said—

> "My Lords, the employers had striven without compromise to prevent shot firers testing in the open. They had done everything that they could to enforce the safety rules. They had been influential in tightening up the regulations imposed on the shot firers personally, they had publicly punished and degraded a shot firer who tested in the open, and they had in consequence faced trouble with the union. They had arranged a system of work and pay designed to encourage the cutting of time and the taking of risks. The two shot firers, George and James, knew all this. In spite of it they deliberately broke the statutory regulations which

were laid on them personally, and together tested in the open. As a result they blew themselves up."[25]

British safety and health legislation has not been afraid to attach personal duties which are enforceable by criminal prosecution to employees at work and section 7 of the 1974 Act is no exception when it provides that it is the duty of every employee while at work:

"(*a*) to take reasonable care for the health and safety of himself and of other persons who may be affected by his acts or omissions at work; and

(*b*) as regards any duty or requirement imposed on his employer or any other person by or under any of the relevant statutory provisions, to co-operate with him so far as is necessary to enable that duty or requirement to be performed or complied with."

The duty under section 7 is twofold: to take reasonable care for the health and safety of himself and others and to co-operate with his employer (or other person) upon whom a duty or requirement has been laid. If an employee or several employees were systematically to break either of the duties in section 7, it would theoretically be possible to serve an improvement notice on him (or them) under section 21 and, possibly, a prohibition notice under section 22 which, however, requires that the notice must relate to certain activities carried on by or under the control "of any person." It might be open to argument that an employee carrying on an activity does not control it within the meaning of section 22. The higher one goes in the echelons of the management structure, the more likely one is to find the necessary degree of "control." The employee's failure to take care or to co-operate with his employer is ultimately a question of fact involving not only standards laid down by law (including standards in any approved code of practice—see section 17, p. 117) but extra-legal standards such as approved practice in an industry including rules and regulations laid down by the employer (although breach of the latter does not necessarily operate as conclusive breach of section 17).

So far as the criminal enforcement of section 7 is concerned, the employee is subject to prosecution under section 33(1) for an offence committed by a person who fails "to discharge a duty to which he is subject by virtue of sections 2 to 7." Justice is often depicted blindfolded, but the inspectorates have indicated that they are not going to rush in with the big stick for every minor infringement.

[25] [1965] A.C. 656, at p. 683.

Examples of cases heard to date, in which employees have been found guilty for breaches of section 7, include a Scottish case where three foundry workers were fined for not wearing the eye protection provided by management,[26] a case where a recalcitrant employee was fined £450 for throwing a hammer at a fellow-worker thereby fracturing the latter's skull,[27] and another case where the driver of a forklift truck was found to have been driving whilst under the influence of drink.[28] For a general discussion of the relationship between health and safety laws and the contract of employment, see p. 22, above.

Failure to comply with section 7 may have less direct effects, as where it leads to a dismissal which is held to be "fair," as in *Marsh* v. *Judge International Housewares Ltd.*,[29] in which an employer dismissed an employee who had refused to have his long hair cut following a warning from the factory inspector that the employer risked prosecution under section 2(1) of the 1974 Act if the employee's hair were to be caught in a machine. A dismissal was similarly held to be fair in *Taylor* v. *Bowater Flexible Packaging Co.*[30] when an employee was eventually dismissed for consistently refusing to wear safety goggles, and in *Simmonds* v. *U.S.A.*[31] in which the employers were held to have acted fairly in dismissing for a first offence a warehouse foreman who had stored incompatible explosives in an explosives store. In *Hudson* v. *Ridge Manufacturing Co. Ltd.*,[32] employers were held liable at common law in respect of potentially dangerous behaviour known to them but which they had failed to deal with either by preventing it or removing its source. In circumstances such as these, an employer is left with little alternative to dismissing the offending employee.

In the U.S.A., section 5(*b*) of the Occupational Safety and Health Act 1970 (29 U.S.C.A. 654) enacts that—

> "Each employee shall comply with occupational safety and health standards and all rules, regulations, and orders issued pursuant to this chapter which are applicable to his own actions and conduct."

In contrast to the British provision, there are no sanctions in the form of citations or penalties against employees who break section 5(*b*). Whilst it might appear that this section is like the

[26] Health and Safety: an Appraisal, W. Simpson, *Managerial Law* 18, VI, 27.
[27] Not always the Employer, Corcoran J., *Health and Safety at Work* (1981) 3, 46.
[28] *Ibid.*
[29] (1977) C.O.I.T. No. 571/57.
[30] (1973) N.I.R.C. (9.5.1973); *cf. Mayhew* v. *Anderson* [1978] I.R.L.R. 10, where the dismissal was held to be unfair.
[31] (1976) C.O.I.T. 561/214.
[32] 1957 2 Q.B. 348.

imperfect law (*lex imperfectus*) known to the Romans, it is likely that the provision will be relevant in determining civil liability just as the shotfirers' breach of regulations in the Shatwell case, quoted above, was relevant in determining the issue of consent to risk. With us, however, there can be no question of section 7 founding a civil claim.[33]

The criminal law leans against vicarious criminal liability so that the employer will not be criminally liable for breach of a duty cast upon the employee (see the case of *Wright* v. *Ford Motor Co. Ltd.*, discussed in connection with section 8, below). If the employer, on the other hand, is guilty of a breach of any of the relevant statutory provisions but that breach is due to the act or default of some other person, such as an employee, the latter person may be proceeded against, whether or not the employer is charged with the breach.[34] A situation may arise in which both employer and employee commit breaches for which each is liable, but this is quite distinct from the concept of vicarious liability in which one is answerable for the act or omission of another person.

DELIBERATE OR RECKLESS INTERFERENCE: SECTION 8

The best health and safety precautions in the last analysis are only as good as the people who operate them. The near-nuclear incident at Three Mile Island in the U.S.A. showed that the weakest link in the safety chain may well be the human link. We have seen above that the Act expects employees to have regard for the safety of themselves and others. Section 8 goes further and deals with those (not necessarily employees) who positively interfere or misuse things. The section, which is framed in categorical terms, states—"No person shall intentionally or recklessly interfere with or misuse anything provided in the interests of health, safety or welfare in pursuance of any of the relevant statutory provisions." Both this and section 6 are not new in placing duties on persons other than those who operate enterprises. Statutory precedents can be found in the now-repealed section 143 of the Factories Act 1961, in section 27 of the Offices, Shops and Railway Premises Act 1963 and, as one might expect, in several provisions of the Mines and Quarries Act 1954.

The *mens rea* (or mental element) for the offence of interference or misuse of things is either intention or recklessness. The

[33] See s.47, discussed below at p. 172.
[34] s.36, discussed below at p. 168.

former signifies foresight of the consequences of what one is doing (or failing to do), with a desire that those consequences should occur; it is essentially a subjective phenomenon, although, like all legal phenomena, it is subject to the vicissitudes of proof. Recklessness involves what has been termed "a wanton disregard for the safety of others," but, more recently, the House of Lords has had occasion to sharpen the definition of recklessness as a mental element in crime.[35] The Law Lords, whilst showing distaste for the "subjective"—"objective" dichotomy, have to some extent given that dichotomy added importance when, in the words of Lord Diplock, they say that a person acts recklessly when he acts "without having given any thought to the possibility of there being any such risk or, having recognised that there was some risk involved, had none the less gone on to take it."[36] Clearly, the requirement that there must have been interference or misuse is, in the present context, an objective requirement. Dealing with a case in which an employee had lifted the panel on a hydraulic mould press and then pressed his leg inadvertently against a starting handle, Hilbery J. construing a statutory provision comparable in some respects to section 8 said—"In my opinion the words 'wilfully interfere or misuse' in the context, more particularly the words 'or misuse,' are intended to mean something more than merely 'touch' or 'misplace.' I think they mean something in the nature of perverse intermeddling with the appliance; and the use of the word 'wilful' in our language can and often does convey that meaning."[37] The employee had, admittedly, intentionally lifted the panel, but this, taken together with the inadvertent pressure on the starting handle, did not amount to wilful interference or misuse, which latter terms should be construed conjunctively. Recklessness is by definition less "wilful" or "deliberate" than "intention" and it is uncertain whether or not section 8 of the present Act would have applied to the employee in the case just discussed. Contrast with this case, the case of *Ginty* v. *Belmont Building Supplies Ltd.*,[38] in which an experienced worker suffered an injury by reason of failing to use crawling boards on an asbestos roof and was held to be disentitled to damages for what was essentially his own fault and also a breach of section 119 of the Factories Act. It is unfortunate that the provisions imposing criminally enforceable

[35] *R.* v. *Caldwell* [1981] 1 All E.R. 961 and *R.* v. *Lawrence* [1981] 1 All E.R. 975.
[36] [1981] 1 All E.R., at p.982.
[37] *Charles* v. *Smith (S) and Sons (Engineering) Ltd.* [1954] 1 W.L.R. 451, 4.
[38] [1959] 1 All E.R. 414.

duties have fallen to be construed in connection with civil claims for compensation.

The section only applies in relation to some thing "provided in the interests of health, safety or welfare in pursuance of any of the relevant statutory provisions." Whether or not a thing is "provided" must be determined in connection with each statutory context in which the word appears; but the broad meaning, as we have seen, is that it is something which is furnished or supplied, notwithstanding that the employer does not give a specific instruction, *e.g.* telling an employee *not* to remove a safety guard.[39] The fact that an employer condones or even approves of a breach does not necessarily exculpate an employee of liability for breach of a duty imposed upon him. As Denning L.J. (as he then was) put it—"it would, I think, be unfortunate if little or no responsibility were to be placed on the workmen themselves."[40] It would seem, however, that a thing cannot be said to be provided if use of that thing is actually forbidden[41] or the thing is not readily available.[42] Although the cases cited above refer to the employee or workman, who will normally be the person guilty of interference or misuse, the section applies to any person who may not necessarily be an employee. An extreme, if unlikely, example would be some industrial saboteur.

Contravention of section 8 is a criminal offence under section 33(1)(*b*) and, as in the case of section 7, criminal liability for that contravention is personal. In *Wright* v. *Ford Motor Co. Ltd.* it was held that the factory occupier was not liable for the act of "X, the unknown man" who had forced open a safety device. Lord Parker C.J. dealing with an argument that the occupier was liable for breach of the strict duty to fence under section 14(1) of the Factories Act preferred to hold that the employer was liable only for contraventions under that Act "other than those in respect of which the sole obligation to perform is imposed on the employee" saying that to find otherwise would "outrage one's sense of the criminal law in regard to vicarious liability of a master for crimes of his servants."[43] As was indicated above in connection with section 7, it is by no means impossible to have criminal liability in both the employer *and* some other person, such as an employee or "outsider," in which each shoulders his own liability.

[39] *Norris* v. *Syndic Manufacturing Co.* [1952] 2 Q.B. 135 and see p.71, above.
[40] S.C. at p.143.
[41] *Murray* v. *Schwachman Ltd.* [1938] 1 KB. 130.
[42] *Finch* v. *Telegraph Construction Co. Ltd.* [1949] 1 All E.R. 452.
[43] [1966] 2 All E.R. 518, 521.

DUTY NOT TO CHARGE EMPLOYEES FOR THINGS DONE OR PROVIDED PURSUANT TO CERTAIN SPECIFIC ENACTMENTS: SECTION 9

Section 9 provides that no employer shall levy or permit to be levied on any employee of his any charge in respect of anything done or provided in pursuance of any specific requirement of the relevant statutory provisions. Clearly, the rationale of this section is to prevent the employer from transferring in whole or in part the cost of health, safety or welfare requirements to his employees. It supersedes section 136 of the Factories Act 1961 (now repealed) which had forbidden deductions by and payments to occupiers of factories. Truck legislation in the nineteenth century sought to end the abuses of the system of "trucking" and "tommy shops" under which employers exploited their workmen either by not paying them the whole of their wages in current coin of the realm (*e.g.* by paying them with vouchers which could be used at the employer's tommy shop) or by making deductions from wages. However, certain deductions were allowed under truck legislation subject to certain safeguards, such as the deductions permissible under section 3 of the Truck Act 1896, which relates to charges for materials, tools[44] or machines, standing room, light, heat, or for or in respect of any other thing to be done or provided by the employer in relation to the work or labour of the workman. However, section 2 of the Hosiery Manufacture (Wages) Act 1874 forbids contracts to stop wages and to charge frame rents (which contracts are to be illegal, null and void).

The prohibition in section 9 refers solely to those things provided in order to comply with the specific requirements of a relevant statutory provision; outside the specific requirements a charge may be levied so far as section 9 is concerned. In *Associated Dairies Ltd.* v. *Hartley*[45] (discussed above at p.66), an improvement notice requiring the provision of protective footwear free of charge was cancelled. The employers had allowed employees working in the warehouse, where roller trucks were used, to purchase safety shoes at cost price. Although an employee had suffered an injury when a truck passed over his unprotected foot, the tribunal was of the opinion (shared by the solicitor to the local authority) that section 9 had no application because it applied only to some specific requirement in the relevant statutory provisions, by which was meant "some statutory instrument or similar regulation enacted *under*

[44] In certain trades the workman is expected to provide his own tools and the employer can pass on the economy of bulk buying to his workman.
[45] [1979] IRLR 171.

the relevant statutory provisions"[46] (*italics supplied*) and not to a general requirement such as that in section 2 of the 1974 Act. The decision itself is based on the question of reasonable practicability (already discussed above) so that the comment on section 9 was not strictly necessary for the decision. Now, whilst it is undoubtedly broadly true that general duties are contained in the statute and specific requirements in the subordinate legislation under that statute, this is not always the case. The statute may descend to the particular[47] and the regulation may ascend to the general (or to the general *and* the particular).[48] With due respect to the tribunal, it seems doubtful that a charge for the fencing of a dangerous part of a machine (under section 14 of the Factories Act 1961) is outside section 9. Section 9 itself refers to "any specific requirement *of* the relevant statutory provisions" and not as the tribunal puts it some subordinate instrument "enacted *under* the relevant statutory provisions." The words "specific requirement" need not mean only a requirement of a detailed and particularised character; it could also mean a requirement which is specified in the sense that section 14 of the Factories Act or section 2 of the 1974 Act are "specific." It would be odd if the legality of charges were to be determined by reference to the distinction between parent Act and subordinate instruments under that Act.

[46] S.C. at p.173.
[47] Several of the requirements in the Factories Act 1961 are particular.
[48] See reg 4, Diving at Work Regulations 1981 (S.I. 1981 No. 399) which in "addition to any specific duty placed on him by these Regulations" goes on to give a general duty.

CHAPTER 4

REGULATIONS, APPROVED CODES AND GUIDANCE NOTES

INTRODUCTION

The Robens Committee envisaged a multi-tiered framework consisting of a general enabling Act supported by a tier of statutes which would be gradually replaced by regulations and a lower tier of non-statutory codes of practice.

New regulations would be kept to a minimum, have a simpler style, be limited to the prescription of general requirements and appear in a form which enabled them to be readily modified.[1] In future, however, much greater reliance would be placed on non-statutory codes and standards—an emphasis which was central to the philosophy of the Report. Notwithstanding that preference, the Committee felt that there might be a need to promulgate approved Codes of Practice where voluntary guidance was lacking or inadequate, and that the standards so set by either type of Code (approved or voluntary) could act as standards of compliance with statutory duties, so that, for example, inspectors could rely on the Codes in addition to statutory provisions to determine compliance or otherwise with standards.

The Committee drew back from advocating that every safety objective should be stipulated in regulations or binding codes. Its members, and in particular, the Chairman, Lord Robens, thought it important that if a business man could meet the requirements of the Act in a better way, then he should be allowed to do so. It was not desired that approved codes should be third-stage law, with voluntary codes as fourth-stage law.

In fact, there have been many more regulations than approved codes, although the Commission has usually made great efforts to explain by means of Guidance Notes (below, p. 118) the regulations that have been made.

REGULATIONS

The General Duties provide the general framework to which more detailed requirements might be related. The Health and

[1] Committee for Safety and Health at Work, Cmnd. 5034, H.M.S.O., 1972, para. 142.

Safety Commission is under a duty to submit proposals for the making of regulations to the Secretary of State[2] who is, in turn, empowered to lay such regulations before Parliament.[3]

Regulations may be made to serve any of thirty different purposes set out in Schedule 3 (see Appendix 1, below).

Regulations may—

A. Prescribe minimum standards and be
   (i) of general application, such as the Safety Representatives and Safety Committees Regulations 1977[4];
   (ii) requirements concerning particular hazards or circumstances, such as the Control of Lead at Work Regulations 1980[5];
   (iii) requirements relating to a particular industry, such as the Diving Operations at Work Regulations 1981.[6]

B. Require that the Health and Safety Commission be notified of circumstances, activities, information etc., such as the Notification of Accidents and Dangerous Occurrences Regulations 1980.[7]

C. Be for licensing or for the granting of prior approval, such as licensing the manufacture of explosives, the certification of plant, the certification of persons to operate high hazard operations and to grant prior permission to undertake high hazard activities.

Any breach of regulations made under the 1974 Act is admissible in evidence in civil proceedings where damage[8] has been caused unless the regulations otherwise provide.

Power to make health and safety regulations

The Secretary of State, the Minister of Agriculture, Fisheries and Food, or the Secretary of State and that Minister acting jointly have power under section 15(1) to make regulations for any of the general purposes of Part I of the Act to be referred to as "health and safety regulations." The generality of this power is not restricted by section 15(2), which permits health and safety regulations for any of the general purposes of Part I to be made for any of the purposes mentioned in Schedule 3. This Schedule (set out in Appendix 1, below), sets out a lengthy list of

[2] Section 11(2)(*d*) Health and Safety at Work etc. Act 1974.
[3] Section 15, Health and Safety at Work etc. Act 11974.
[4] S.I. 1977 No. 500.
[5] S.I. 1980 No. 1248.
[6] S.I. 1981 No. 399.
[7] S.I. 1980 No. 804.
[8] Damage here includes the death of, or injury to, any person (including any disease and any impairment of a person's physical or mental condition). See section 47 of the Health and Safety at Work etc. Act 1974, below p. 173.

matters on which health and safety regulations may be made. The Schedule covers an extremely wide range of matters extending to welfare arrangements, the management of animals and the making of bye-laws.

Section 15(3) enables health and safety regulations to—

(a) repeal or modify any of the existing statutory provisions;
(b) exclude or modify in relation to any specified class of case any of the provisions of section 2 to 9 or any of the existing statutory provisions;
(c) make a specified authority or class of authorities responsible, to such extent as may be specified, for the enforcement of any of the relevant statutory provisions.

These are radical powers in that they permit Acts of Parliament to be repealed or modified and a latter-day "dispensing power" to be exercised in relation to section 2 to 9 of the parent Act itself. Whatever the constitutional purist may say, it should be remembered that the objective of the new system is to replace the existing corpus of statutes and subordinate legislation by health and safety regulations—an objective on which the realism of the early years is now tempered by a more pragmatic realism. A precedent for the power to alter the terms of a statute or the application of a statute by means of subordinate legislation is afforded by section 76 of the Factories Act 1961, which conferred a wide ministerial power to make special regulations under that Act and also to modify or extend certain provisions of that Act. Thus, the power to modify the statutory provisions enabled special regulations to be made in substitution for those provisions as is expressly stated in some regulations such as the Abrasive Wheels Regulations 1970 (S.I. 1970, No. 535), which in relation to abrasive wheels are in substitution for section 14(1) of the Factories Act 1961.[9] Where the regulations are silent on the matter, their terms will generally be held to be in substitution for the provisions of the Act so far as the subject-matter of the regulations is concerned.[10] In respect of matters not expressly provided for in the regulations, the general obligations of the Act remain in force.[11] Whether or not the power to modify the provisions of the Act included the power to abrogate an absolute duty imposed by that Act, without the substitution of other requirements, is a question on which differing views have been expressed *obiter* in the House of Lords. There is no doubt under section 15(3) of the

[9] Regs. 2(2) and 3(2).
[10] *Miller* v. *William Boothman and Sons Ltd.* [1944] K.B. 337; *Automatic Woodturning Co. Ltd.* v. *Stringer* [1957] A.C. 544.
[11] *Benn* v. *Kamm and Co. Ltd.* [1952] 2 Q.B. 127; *Dickson* v. *Flack* [1953] 2 Q.B. 464.

1974 Act which unequivocally allows of the repeal or modification of existing statutory provisions and dispensations in relation to any specified class of case with regard to the general duties in section 2 to 9 or any of the existing statutory provisions.

By virtue of section 15(4), health and safety regulations may—

(a) impose requirements by reference to the approval of the HSC or any other specified body or person;
(b) provide for references in the regulations to any specified document to operate as references to that document as revised or re-issued from time to time.

Section 15(5) permits health and safety regulations to—

(a) provide (either unconditionally or subject to conditions, and with or without limit of time) for exemptions from any requirement or prohibition imposed by or under any of the relevant statutory provisions;
(b) enable exemptions from any requirement or prohibition imposed by or under any of the relevant statutory provisions to be granted (either unconditionally or subject to conditions, and with or without limit of time) by any specified person or by any person authorised in that behalf by a specified authority.

Section 15(6) continues the litany with health and safety regulations which may—

(a) specify the persons or classes of persons who, in the event of a contravention of a requirement or prohibition imposed by or under the regulations, are to be guilty of an offence whether in addition to or to the exclusion of other persons or classes of persons;
(b) provide for any specified defence to be available in proceedings for any offence under the relevant statutory provisions either generally or in specified circumstances;
(c) exclude proceedings on indictment in relation to offences consisting of a contravention of a requirement or prohibition imposed by or under any of the existing statutory provisions, sections 2 to 9 or health and safety regulations;
(d) restrict the punishments which can be imposed in respect of any such offence as is mentioned in (c) above (other than the maximum fine on conviction on indictment).

Section 15(7) enables health and safety regulations to make provision for enabling offences under any of the relevant statutory provisions to be treated as having been committed at any specified place for the purpose of bringing any such offence within the field of responsibility of any enforcing authority or conferring jurisdiction on any court to entertain proceedings for any such offence. This subsection is, however, without prejudice to section 35 of the Act (which deals with the venue of offences under any of the relevant statutory provisions committed in connection with plant or substance). Section 15(8) permits of considerable refinement by allowing health and safety regulations to apply to particular circumstances only (*e.g.* regulations applying to particular premises only).

We have seen above that the operation of the 1974 Act has been extended outside Great Britain by means of an Order in Council.[12] Section 15(9) provides that if an Order in Council is made under section 84(3), applying section 15 to or in relation to persons, premises or work outside Great Britain, then, notwithstanding the Order, health and safety regulations shall not apply to or in relation to aircraft in flight, vessels, hovercraft or offshore installations outside Great Britain, or to persons at work outside Great Britain in connection with submarine cables or submarine pipelines, except in so far as the regulations expressly so provide.

Regulations under relevant statutory provisions: section 50

Powers to make regulations under any of the relevant statutory provisions exerciseable by the Secretary of State, the Minister of Agriculture, Fisheries and Food, or both of them acting jointly, may be exercised either—

(a) to give effect (with or without modifications) to proposals submitted by HSC under section 11(2)(*d*)—the normal procedure, or
(b) independently, *i.e.* without proposals from the HSC—in which case there must be consultation with HSC and such other bodies as appear to him to be appropriate.

There must also be consultation with HSC before the making of regulations where it is proposed to modify HSC proposals under (a) above.

In the ordinary course of events, regulations will be proposed by HSC under section 11(2)(*d*); but before submitting them it must consult with any government department or other body that appears to it to be appropriate, and, in particular to consult any body representing local authorities where it is proposed to

[12] See Chapter 1, p. 37.

make local authorities the enforcing authorities or the National Radiological Protection Board in the case of proposals to make regulations relating to electro-magnetic radiations. It must also consult with such government departments and other bodies, if any, as, in relation to any matter dealt with in the proposals, it is required to consult under section 50(3) by virtue of any directions given to it by the Secretary of State (see section 12 on control of HSC by the Secretary of State).

General power to repeal or modify Acts and instruments

Section 80 confers a general power on the Secretary of State, the Minister of Agriculture, Fisheries and Food, (or the two acting jointly) to repeal or modify any of the following provisions (other than the relevant statutory provisions) where to do so is expedient in consequence of Part I of the Act. The provisions in question are those which—

(a) are contained in the 1974 Act or pre-1974 Acts; or
(b) are contained in any regulations, order or other instrument of a legislative character which was made under an Act prior to the 1974 Act; or
(c) apply, exclude or for any other purpose refer to any of the relevant statutory provisions and which are contained in any Act not falling within (a) above or in any regulations, order or other instrument of a legislative character which is made under an Act but does not fall within (b) above.

The general power to repeal or modify, described above, has been extended to provisions in the Employment Protection (Consolidation) Act 1978 which re-enact provisions in the Redundancy Payments Act 1965, the Contracts of Employment Act 1972 or the Trade Union and Labour Relations Act 1974 in the same way as the general power applies to provisions contained in Acts passed before or in the same session as the 1974 Act. Without prejudice to the generality of the general power, regulations making modifications may modify the enforcement provisions to which section 80 applies (including the appointment of persons for the purpose of such enforcement, and the powers of persons so appointed). By virtue of section 82(1), the term "modifications" includes additions, omissions and amendments.

Parliamentary control

Regulations made under Parts I, II and IV of the 1974 Act take the form of statutory instruments which are subject to annulment in pursuance of a resolution of either House of Parliament—the so-called "negative procedure." This is provided for in section 82(*b*) which thereby ensures some degree of Parliamentary control. References to Parliamentary control in effect signifies parliamentary interest by a small number of M.P.'s from the different political parties who have an especial interest in health and safety at work. Important as that subject is, it does not seem to have the drawing power of more emotive subjects such as the death penalty or abortion. It may be that the entrusting of public functions to regulatory bodies such as HSC and HSE tends on the whole to weaken Parliamentary control of those bodies. An unusual feature of the 1974 Act is the complexity in the regulation-making powers. We have seen that the main power is to be found in section 15 which regulates the making of health and safety regulations; but other powers lie scattered about the Act, namely, section 43 (financial provisions), section 50 (which, as we have seen, deals with regulations under the relevant statutory provisions) and section 80 (containing the general power discussed above). Finally, section 84, which, as we have seen,[14] has already been used to extend the operation of the Act outside Great Britain, provides for Orders in Council which are subject to annulment in pursuance of a resolution of either House of Parliament. Both statutory instruments and Orders in Council under the Act may be drafted in flexible terms so as to make different provision for different circumstances or cases.

There has predictably been criticism of health and safety legislation emanating from the EEC, mainly based on the lack of prior parliamentary discussion. This criticism is not peculiar to health and safety legislation but raises more fundamental questions concerning Parliamentary sovereignty. As we were the first country to industrialise, it is not surprising that for a long time our protective legislation was far in advance of that in countries whose industrialisation took place later than our own. Now, however, there is no doubt that a major catalyst for reform is to be found in the "tide" of EEC regulation of health and safety both for workers and consumers. The impact of the EEC (and of the ILO) is discussed below in Chapter 12. The fears that Parliament is being bypassed, stem from a feeling that we are witnessing the rise of legislation by ministerial decree, rather like the exercise of the *pouvoir réglementaire* in France where parliamentary accountability is less strong than it is with us. If in constitutional theory Parliament can do anything (a view

[13] See Chapter 1, p. 37.

now stated with increasing hesitancy) it does not follow that it must therefore do everything.

A back-up to Parliamentary control is to be found in the Joint Select Committee on Statutory Instruments which in 1973 replaced the separate Committee in the Commons and Lords formed to scrutinise delegated legislation consisting of statutory instruments of a general character and certain other instruments which are laid or laid in draft before the House and are subject to the negative or affirmative resolution procedure. The Joint Committee can draw attention to an instrument on certain grounds set out in its terms of reference, such as doubts concerning the vires of the instrument or the unusual or unexpected use made of the powers given by the parent Act. The Chairman of the Joint Select Committee, Mr. Bob Cryer, has drawn attention to features of regulations in parliamentary proceedings involving those regulations, *e.g.* to a regulation increasing fees for medical examinations which, he alleged, eroded the primary duty to maintain the Employment Medical Advisory Service.[14] Mr. Cryer considered that whilst the Act "allows the Minister to do a wide range of things, far wider than I regard as reasonable. . . . Nevertheless, it does not allow him to erode the provision of the EMAS."[15]

Cost of regulations

Since 1974, when the Act received the Royal Assent, the oil crisis and the supervening world recession have placed a brake upon the introduction of regulations which would add considerably to costs. The HSC is taking steps to implement Sir Leo Pliatzky's recommendation that it should publish appraisals of the cost to industry and employers of the measures it proposes, together with estimates of the gains from those measures.[16] This is all part of the greater emphasis upon cost-benefit analysis in an age which has abandoned the rather more careless attitudes of the Sixties. It is significant that we still await details of prescribed information which employers and the self-employed must give to persons other than their employees under section 3(3) and, similarly, we still await the regulations prescribing information to be included in directors' reports for which section 79 makes provisions.[17] It may well be that the impediments are to be found in the preference for advance by means of consensus

[14] Parl. Deb. H.C., June 8, 1981, col. 232: see also, for example, February 26, 1980, col. 1323 (term requiring elucidation); April 28, 1980 (*ultra vires* and warrants to inspectors); July 29, 1980, col. 1428 (creation of criminal offence and alleged dilution of stringency of notification of accidents requirements).

[15] Parl. Deb. H.C., June 8, 1981, col. 232: see also April 27, 1981, col. 620 on the Health and Safety (Diving) Operations; February 26, 1980, col. 1323; April 28, 1980, col. 322.

[16] Parl. Deb. H.C., May 8, 1980, col. 203: see also March 11, 1977, col. 700.

[17] See below, p. 186.

between the two "sides" of industry, or in the resources and priorities of HSC/HSE, or in some other factor divorced from cost; but cost is increasingly likely to be one of the considerations affecting the introduction of implementary regulations such as those just mentioned as well as future regulations.

APPROVED CODES OF PRACTICE

Section 16 of the 1974 Act provides for the issue and approval of codes of practice "for the purposes of providing practical guidance" in relation to the requirements of sections 2–7, or health and safety regulations under the Act. By virtue of section 17, a failure on the part of any person to observe any provision of an approved code of practice does not of itself render that person liable to civil or criminal proceedings. Whilst such a code does not have direct effect, any provision in such a code which appears to a court to be relevant to an alleged contravention of a requirement or prohibition is admissible in evidence. If it is proved that there was at any material time a failure to observe any provision of a code which the court considers to be relevant for the prosecution to prove in order to establish a contravention of a requirement or prohibition, that matter will be taken as proved unless the court is satisfied that the requirement or prohibition was in respect of that matter complied with otherwise than by way of observance of that provision of the code.

To illustrate the operation of this latter provision, let us assume that new Lighting Regulations have been made under section 15 of the 1974 Act and that these provide, *inter alia*, that an employer is obliged to provide lighting in a window-less office of not less than thirty lumens per square foot. The relevant code of practice indicates that lighting of a sufficient standard will have been provided if gas or electric incandescent or discharge lamps are situated so as to avoid glare, direct or reflected, sharp contrasts or deep shadows. Let us further assume that employer X fails to provide gas or electric lighting of the stipulated variety but does supply petrol vapour lamps for his offices which are situated in a remote rural area without gas or electrical supplies. If prosecuted, it will be for the prosecution to prove a breach of the code (relating to gas or electric lamps). If this is done, it will be for the defendant X to prove on the balance of probabilities that the provision of petrol vapour lamps gives sufficient and suitable lighting with not less than thirty lumens per square foot and subject to the accompanying requirements as to glare and so forth. Improvement and Prohibition Notices may be framed by reference to an approved code of practice and the approved code of practice thus becomes a part of the notice. See section 23(2)(*a*) of the Health and Safety at Work etc. Act 1974.

Before approval can be given by the Health and Safety Commission it is required to consult appropriate Government Departments and other interested parties. The Commission believes that in order for standards to be effective they must be agreed by all parties, including manufacturers, employers and workers. Because this procedure is necessarily lengthy, the number of approved Codes of Practice had, at the time of writing, not exceeded six (listed in Appendix 2).

The legal significance of approved codes of practice, as has been seen, closely resembles that of the Highway Code in relation to the Road Traffic Act 1972 and the "deemed to satisfy" provisions of the Building Regulations 1976. There are however clear differences. A failure on the part of a person to observe a provision of the Highway Code will not render a person liable to criminal proceedings *per se*, although it may be relied upon by any party to the proceedings as tending to establish or to negative any liability which is in question in those proceedings, although like the Highway Code, approved codes of practice are admissible in evidence in both criminal and civil proceedings.[18]

> "A breach of the Highway Code might be relied on as tending to establish, in civil proceedings, liability on the part of the person in breach but a breach creates no presumption of negligence calling for an explanation, still less a presumption of negligence making a real contribution to causing an accident or injury; it is merely one of the circumstances on which one party was entitled to rely in establishing the negligence of the other."

Powell v. *Phillips* [1972] 3 All E.R. 864.

Whilst in relation to "deemed to satisfy" provisions compliance with that standard is one way of satisfactorily fulfilling the legal obligation, it is axiomatic that such a provision does not either require or ensure exclusive conformity with a particular standard.

GUIDANCE NOTES

The lowest tier is made up of Guidance Notes issued by the Health and Safety Commission and the Health and Safety

[18] See section 37 of the Road Traffic Act 1972.

Executive, by Industry Advisory Committees[19] and by Industry itself. Guidance Notes issued by the Health and Safety Commission are authoritative in so far as they carry the Commission's *imprimatur*; but neither they nor Guidance Notes issued by the Health and Safety Executive, have been invested with any legal significance. Their aim is to be helpful and to assist employers and others to comply with the law. The same is true *mutatis mutandis* in relation to guidance issued by the Industry Advisory Committees and to Industry-generated notes of guidance.

STANDARDS

An important development since the enactment of the 1974 Act has been the increasing reliance of industry and the inspectorates on technical standards. Some of these standards, such as those issued by the British Standards Institution, may not have been formulated with safety and health at work as the main consideration. Sometimes a standard such as a BSI standard may be incorporated in Regulations, such as the Safety Signs Regulations (S.I. 1980 No. 1471) which require compliance with BS 5378: Part I: 1980. Other examples of BSI standards may be found, not only in Regulations, but also in Commission Guidance Notes, including Guidance Notes issued to local authorities which are enforcing authorities. Such standards may be used quite apart from their incorporation in Regulations, Codes of Practice or Guidance Notes, and the same applies to standards published by HSE itself (including those contained in the reports of Industry Advisory Committees) or arrived at as voluntary codes agreed by employers and trade unions for their own industry. Section 16 of the 1974 Act permits HSC to approve standards whether prepared by it or not.

Because of a wish expressed by some manufacturers and users of equipment for more guidance and the need to use more fully the internationally authoritative bodies based in the United Kingdom, the Health and Safety Commission has approved a plan for the development by the British Standards Institution of ten product standards relating to topics ranging from industrial safety helmets to flameproof electrical equipment.[20]

The difficulties in seeking to use British Standards on a large scale would seem to stem from two sources. First, not all British Standards would be suitable; some contain references to other

[19] Of the Health and Safety Commission.
[20] Health and Safety Commission, Plan of Work 1981–1982, H.M.S.O. 1981, pp. 19–20. This programme will be jointly supervised by the Department of Industry.

standards, others are vague on important health and safety matters and yet others are closely aligned to international standards making amendment difficult. Second, approved codes of practice have a special legal significance under the Health and Safety at Work etc. Act 1974 and care will have to be exercised that any amendments made by the British Standards Institution in relation to an adopted standard have undergone the appropriate level of scrutiny by those whom the Health and Safety Commission is bound to consult before approving new codes of practice.

From a legal standpoint, the use of general standards such as that in section 2 of the Act relating to plant and safe systems, enables the law to be applied widely and flexibly. A wide interpretation of the notion of a "system of work" can be given and the standard of care in providing and maintaining such a system can be kept in line with contemporary standards. General standards have also been used in regulations made under the 1974 Act which, it is hoped, will enable the joint approach used in the Safety Representatives and Safety Committees Regulations 1977 to be supported by what the HSC has termed "a simple, flexible and acceptable system of legislation."[21] The HSC has indicated that—"Where regulatory controls are needed they are being drafted wherever possible in terms of objectives so that employers and others are provided with adequate flexibility on how to achieve them, as set out in approved codes of practice and guidance."[22] This is in line with the principle of self-regulation commended by the Robens Committee. It would be impossible to provide specific and detailed guidance in regulations on all of the thirty subject-areas set out in Schedule 3 to the 1974 Act and to back these up with an ever-increasing army of inspectors who could use their enforcement powers. It is interesting to note that the Advisory Committee on Major Hazards, in their Second Report, consider that the proposed notification and hazard survey regulations should require a company to identify its own problems, to determine the standards, systems and priorities which are to apply and to demonstrate to HSE the conclusions which have been arrived at and the solutions which are proposed. The Committee considers that—

> "Inductive regulations of this kind would have many attractions. In particular there is no need to limit in advance the range of installations to which they might be applied. The existing enforcement procedures by improve-

[21] Report 1980–81, para. 13.
[22] *Ibid.*, para. 63.

ment and prohibition notices would automatically apply and they could be made to affect existing installations without difficulty."

The standard-setting function is also a central function of the American Occupational Safety and Health Administration (OSHA) set up under the Occupational Safety and Health Act 1970 which requires both employer and employee to comply with standards promulgated under the Act. Initially, OSHA promulgated existing national consensus standards and established federal standards; but since then it has turned its attention to promulgating its own standards, particularly in the field of toxic or carcinogenic substances. It will be seen therefore that the borrowing of consensual norms in industry and the attaching of legal sanctions thereto is not confined to this country.

One of the problems at the frontier of science is that of converting a known risk into some regulatory standard which avoids the Scylla of undue alarm and the Charybdis of undue complacency. It may be "known" that a substance is a health risk without knowing what is a safe limit for that substance. As is not unusual in this imperfect world, we sometimes have to do the best we can with imperfect knowledge, avoiding the "absolutist" solution of a "zero risk" where this would involve disproportionate economic costs and social dislocation. The recent American case of *Industrial Union Department, AFL-CIO* v. *American Petroleum Institute,*[23] in which the Supreme Court affirmed a finding of invalidity in relation to a reduced benzine limit promulgated by the Secretary of Labor at the behest of the Occupational Safety and Health Administration (OSHA) illustrates the problem. The American Occupational Safety and Health Act 1970 uses a dual approach by imposing upon each employer the duty to furnish to each of his employees employment and a place of employment which are free from recognised hazards that are causing or are likely to cause death or serious physical harm to his employees (the "general duty") and a duty to comply with occupational safety and health standards (the "particular duty"). The OSHA standard which was struck down by the Courts was one which reduced the threshold limit for airborne concentrations of benzine from 10 ppm to 1 ppm, a reduction based upon scientific reports, including a recommendation from the National Institute for Occupational Safety and Health. The standard was declared to be invalid because (a) there was no evidence as to the significant risks which could be eliminated by the change and (b) there was no evidence on the

[23] 65 L.Ed. 2d., 1010; 448 US 607a, 100 S Ct 2844.

relationship between the expected benefits of the change and the costs which it would impose. The Supreme Court was not saying that the OSHA rationale for the change was wrong, only that it had not been supported by sufficient evidence on the part of OSHA who had not discharged the required burden of proof to justify the change. Mr. Justice Rehnquist, concurring in the plurality opinion of the Court, conceded that "it may be true, as suggested by Mr Justice Marshall, that the Act as a whole expresses a distinct preference for safety over dollars. But that expression of preference, as I read it, falls far short of the proposition that the Secretary must eliminate marginal or insignificant risks of material harm right down to an industry's breaking point."[24] The decision typifies the problem in converting scientific knowledge of *some* risk into a standard of *acceptable* risk. As de Tocqueville noted on his travels in the newly-formed Union, there are few matters in America which are not submitted for the decision of the Courts. The American Act specifically confers upon any person who may be adversely affected by a standard issued under the Act the right to challenge the validity of that standard in the Courts. In its early days, OSHA made use of existing consensus standards and established federal standards; but more recently it has turned its attention to promulgating its own standards, particularly in the field of toxic and carcinogenic substances. Indeed, from this side of the Atlantic, OSHA seems generally to have had a rough passage in its dealings with the law.

Quite apart from this standard-setting process, inspectors in their day-to-day work constantly refer employers and others to technical standards and would clearly be embarrassed if they were called upon to prosecute those who had relied on their advice concerning a standard which, it is alleged, does not discharge the legal duty. The tribunals and courts are not obliged to accept the *ipse dixit* of an inspector on the acceptability of a technical standard (other than one contained in legislative or quasi-legislative form) in concretising a legal standard; but the fact remains that health and safety law, like the criminal law, is not self-enforcing. Individuals—inspectors or police officers—are the persons who normally set the law in motion and this statutory discretion confers upon them great *de facto* normative authority.

[24] 65 L.Ed. 2d., 1060.

CHAPTER 5

ENFORCEMENT BY ADMINISTRATIVE SANCTIONS

By virtue of section 18, the duty to make adequate arrangements for the enforcement of Part I of the 1974 Act and the existing statutory provisions falls upon HSE, although as we have seen, the Secretary of State has made regulations transferring certain health and safety functions to local authorities—mainly their pre-existing statutory functions as enlarged by the new powers given by the 1974 Act.[1] The local authorities are assisted by HSC in the form of a flow of circulars, guidance notes and information to enable them to exercise their duties under the 1974 Act and, also, by the HSE in the form of technical back-up services such as the provision of laboratory services or expert advice.

The word "enforcement" conjures up the Austinian model of punitive sanctions. In "enforcing" the law, the enforcing authorities operate (1) by advice and persuasion (2) by the use of the administrative sanctions described in this Chapter, and (3) by resort to criminal prosecution (which forms the subject-matter of the next Chapter). The fact that the law is largely concerned with the pathology of social relationships should not obscure the healthy physiological side of health and safety at work in which the inspectorates and those subject to their ministrations if not joined in a symbiotic union do often work closely and harmoniously together. The achievements of the inspectorates using techniques of persuasion and advice although unsung outweigh achievements by the more dramatic and more publicised use of sanctions.

In this Chapter we concentrate upon enforcement policy, the appointment of inspectors, their powers and indemnification, and the considerable enforcement powers (mainly by means of improvement notices and prohibition notices) given to inspectors by the 1974 Act which are exerciseable not only in relation to the 1974 Act itself, but also in relation to the existing statutory provisions. The most extreme form of enforcement—the criminal prosecution—is to be found in the next Chapter.

Enforcement Policy: Health and Safety Executive

The principal tasks of the six inspectorates under the aegis of the Health and Safety Executive are:

[1] See Chapter 2 on local authorities, pp. 53 *et seq.* On liaison, see below, p. 126.

(a) the inspection of work activities at workplaces;
(b) providing advice to employers, employees and members of the public;
(c) investigating accidents and cases of occupational ill-health;
(d) remedying complaints from employees and members of the public;
(e) the legal enforcement of health and safety legislation.[2]

The oldest and largest of these inspectorates, H.M. Factory Inspectorate has the most highly developed enforcement policy. Indeed, three of the Inspectorates, H.M. Alkali and Clean Air Inspectorate, H.M. Mines and Quarries Inspectorate, and H.M. Nuclear Installations Inspectorate have demonstrated a reluctance to prosecute or use the enforcement notice procedures in the Act.

In August 1976 the Health and Safety Commission considered the inspection priorities of the Factory Inspectorate. Their aim was to:

(1) Ensure that the efforts of inspectors should be concentrated on premises or activities which are likely to be most productive in furthering the objectives of the 1974 Act.
(2) Ensure that a satisfactory balance is maintained between:
    (i) Special investigations and inquiries.
    (ii) Reactive visits to investigate accidents, complaints, etc.
    (iii) The programme of basic inspection prepared locally.
    (iv) Provision of information and advice.

It was agreed that premises selected for basic inspection should be listed taking into account the following factors:

(a) the present standards of health, safety and welfare in the workplace;
(b) the size and nature of the worst problem that could arise, whether in terms of a single incident or a long-term health hazard, considered in terms both of employees and the public at work;
(c) management's ability to maintain acceptable standards;
(d) the length of time since the previous inspection and the consequent increased possibility that standards have deteriorated or that new hazards have emerged.

[2] See Health and Safety Commission, Plan of Work 1981–82 and onwards, at page 9.

H.M. Factory Inspectorate is responsible for enforcing health and safety legislation in 750,000 workplaces. By the end of 1985, it is planned that all premises and employers will be selected for inspection on the basis outlined above. It is proposed, however, to continue to devote about fifteen per cent of inspection resources to the construction industry. This industry has the special problems of a high accident rate and a constantly changing workplace, allied to the difficulties in dealing with a multitude of employees of varying size and commitment to health and safety. H.M. Agriculture Inspectorate is responsible for enforcement and advice in 250,000 workplaces. In addition to farms, they include agricultural contractors, forestry workers and some new entrant work activities (zoos and safari parks). Particular areas of concern include farms which employ labour, and the investigation of particular hazards such as asbestos and dust. H.M. Mines and Quarries Inspectorate is responsible for inspections in mines worked by the National Coal Board, coal mines worked by private owners under licence from the Board, and Quarries. The numbers killed or seriously injured in this industry have not significantly decreased since 1974. It has been decided to increase the number of inspectors to restore the inspection effort to the 1974 level.

The number of inspectors in H.M. Nuclear Installations Inspectorate is to be substantially increased because of the potentially serious effect of a major incident and of increased public apprehensions on this score. At least two incidents, in which life and health have been endangered, occurred at Windscale during its first 25 years of operation. In addition, there were 25 lesser incidents.[3] Since 1976, when stricter reporting requirements were introduced, there have been 30 reported incidents per year.[4] Perhaps, the most serious accident to occur at Windscale was in 1957 when eleven tons of uranium caught fire and a disaster was only narrowly averted.[5]

The remaining inspectorates, H.M. Alkali and Clean Air Inspectorate and H.M. Explosives Inspectorate, are small and specialised. The former is engaged in a selective area of pollution control and oversees 2000 Alkali and Scheduled Works embracing 3000 installed processes; whilst the latter inspects and licenses explosive factories, investigates accidents involving explosives; deals with the control of imports of explosives, liaises with local authorities and the public on the storage and illegal manufacture of explosives and provides advice on explosives and other hazardous materials.

[3] Parl. Deb. H.C., January 10, 1977, col. 347 (Written Answers).
[4] Windscale: *The Management of Safety*, F.R. Charlesworth *et al.*, HSE 1981.
[5] *Nuclear Energy: the serious doubts that put our future at risk*, Sir Martin Ryle, The Times, December 14, 1976, p.14.

Local Authorities

Local authorities are responsible for enforcing health and safety legislation in some 600,000 premises and since the Health and Safety at Work etc. Act 1974 came into force there has been a full exchange of ideas at national and local level. At national level it was agreed by the Health and Safety Commission and the Local Authority Associations to set up a Committee comprising officers of the Health and Safety Executive and representatives from the Scottish and English and Welsh Local Authority Associations. This committee which is known as the Health and Safety Executive Local Authority Enforcement Liaison Committee has the following terms of reference: "To discuss informally matters of mutual interest between the local authority associations and the Health and Safety Executive concerning the implementation and enforcement of the Health and Safety at Work etc. Act 1974, and to exchange information on standards of inspection and enforcement, training of staff, liaison between local authorities and the field organisation of the Health and Safety Executive and on standards of recording and reporting information for statistical purposes." At the local level, liaison is maintained between local authorities and the Health and Safety Executive by means of the Enforcement Liaison Service.[6] In each Area Office of the Health and Safety Executive one or more senior H.M. Inspectors of Factories have been made responsible for keeping in close touch with local authority officers in their Area who have been appointed to enforce the Health and Safety at Work etc. Act 1974.

Section 18(4)(*b*) Health and Safety at Work etc. Act 1974 requires every local authority to perform its enforcement duties in accordance with guidance issued by the Commission. A number of memoranda have been issued to local authorities by the Commission to facilitate the adoption of consistent enforcement policies. One such circular, entitled *Enforcement Policy and Methods of Inspection*, advocates the adoption of a system of priorities to ensure that inspection activities are directed to the areas of greatest need. Such a system should, it states, be directed towards:

(1) the elimination of the more serious hazards and risks to health.

(2) a quick response to complaints and requests for advice which either indicate the occurrence of an immediate hazard or, in the case of persons employed, that the

[6] Replacing the Central Advisory Service established under the Offices Shops and Railway Premises Act 1963 by the Department of Employment.

workpeople have been unable to resolve the issue in direct discussion with management or through safety representatives;

(3) the development of the use of notice procedures in accordance with previously issued guidance.

In addition local authorities should exercise regular surveillance over all activities within their jurisdiction even though some of these may be relatively low risk and carried out only on a small scale.

Appointment of Inspectors

Section 19 authorises every enforcing authority to appoint as inspectors (under whatever title it may from time to time determine) such persons having suitable qualifications as it thinks necessary for carrying into effect the relevant statutory provisions within his field of responsibility; it may also terminate any appointment under this section. Every appointment of a person as an inspector under this section shall be made by an instrument in writing specifying which of the powers conferred on inspectors by the relevant statutory provisions are to be exercisable by the person appointed; and an inspector in right of his appointment is entitled only to (a) exercise such of those powers as are so specified, and (b) exercise the powers so specified within the field of responsibility of the authority which appointed him. The inspector's powers may be varied by the enforcing authority which appointed him. An inspector must, if so required when exercising or seeking to exercise any power conferred on him by any of the relevant statutory provisions, produce his instrument of appointment or a duly authenticated copy thereof.

The authority of inspectors has been a source of controversy both in Parliament and in the Courts. It has been disclosed in Parliament that the practice is to issue inspectors with an instrument of appointment together with a certificate of appointment signed by the Director General of HSE.[7] Some concern has been expressed concerning the alleged failure to identify, on the warrant issued to an inspector, the particular statutory provisions which he is authorised to implement.[8] The

[7] Parl. Deb. H.C. (Written Answers), July 20, 1976, col. 415.

[8] Parl. Deb. H.C. June 23, 1980 cols. 191–192: see Written Answers, June 5, 1980, col. 843; January 19, 1981, col. 46.

majority of HSE inspectors hold full warrants[9] specifying that the inspector can enforce powers under the relevant statutory provisions which, *ex facie*, would allow an inspector from the nuclear inspectorate to enforce regulations in a coal mine or on a farm.[10] As we have seen, the notion of a unified health and safety inspectorate is far from realisation, and we are informed that—"Administrative arrangements are made by the Health and Safety Executive so that a member of one inspectorate does not do the work of another inspectorate,"[10] an assurance which does not appear to have mollified critics who, accepting that an inspector can be required to produce his authority, answer that the authority produced does not specify such "administrative" limitations of authority. The government has taken the view, notwithstanding comments by the Joint Select Committee on Statutory Instruments that section 19 does not require administrative limitations of authority to be set forth on the warrant, a view which has been said to condone a practice which is *ultra vires*.[11] Section 20 (discussed below) sets out the powers of inspectors to carry into effect any of the relevant statutory provisions; but the critics consider that the need to specify on warrants which of the powers conferred on inspectors by the relevant statutory provisions are to be exercised by those inspectors extends beyond section 20 and requires the relevant statutory provisions themselves to be identified. This, is, at the least, a tenable view although admittedly it would render the preparation of warrants an extremely complicated and difficult process. Distinctions are, however, made in the powers conferred upon inspectors, inasmuch as assistant inspectors of factories are not normally given power to prosecute under section 39 of the Act (which requires that an inspector must be authorised by his enforcing authority to conduct prosecutions in the magistrates' courts for an offence under any of the relevant statutory provisions), or to issue improvement and prohibition notices under sections 21 and 22. A paragraph in Schedule 2 (para. 20(3)) which deals with instruments appointing persons to the Executive was held to be directory and not mandatory, and it was not necessary for the inspector to show that those who had appointed him had themselves been validly appointed under the

[9] It was reported (Written Answers, February 1, 1980, col. 803) that the following categories of inspectors held full warrants—

Agricultural inspectors—all grades except assistant agricultural inspectors.

Alkali and Clean Air Inspectors—all inspectorial grades.

Factory Inspectors—all general and specialist inspectorial grades with the exception of inspectors of factories (fire), fire surveyors and assistant inspectors of factories.

Mines and Quarries—all inspectorial grades.

Nuclear Installations—all inspectorial grades.

[10] Parl. Deb. H.C. (Written Answers), June 5, 1980, col. 843.

[11] Parl. Deb., H.C., June 23, 1980, col. 191.

paragraph. The relief with which this conclusion was reached can be judged from the 194 cases of prosecution by the HMFI which had been adjourned pending the outcome of the appeal in the Campbell case.[12] That case does not relieve inspectors of the need to show their authority, but places reasonable limits on their obligation to show that authority by an unbroken *catena* leading back to HSE and HSC.

Some consternation was caused in the early years of the operation of the Act by challenges brought by persons against whom informations had been preferred relating to offences to be tried in the magistrates' courts, of which the most widely publicised case was that in *Campbell* v. *Wallsend Slipway and Engineering Co. Ltd.*[13] In that case an HSE inspector had laid informations against a company for alleged failure to comply with asbestos regulations. The company pleaded not guilty to the charge, but requested the justices to try two preliminary points before dealing with the substantive charge, *viz.* the legal competence of the inspector to prosecute under section 38 and the inspector's authority to prosecute before a magistrates' court under section 39. Reversing the decision of the magistrates on these preliminary points, the Divisional Court held that section 38 is a limiting section cutting down the general right of anyone to lay an information and is not an empowering section; but that, in any case, it was not enough to challenge the appointment or powers of the inspector, or the appointment of the enforcing authority which has appointed him, without leading evidence or rebutting the presumption of regularity—*omnia praesumuntur rite esse acta.* The inspector gave evidence to the justices that he was an inspector and produced supporting documents, two of which were signed by a director of HSE (with an HSC document vouching for the appointment of the director).

Powers of Inspectors

Section 20 confers upon inspectors who have been validly appointed under section 19, sweeping summary powers[14]; summary in the sense that they can be exercised without recourse to the courts although their exercise can be questioned in the courts. Section 20 sets forth the powers which may be exercised by an inspector for the purpose of carrying into effect any of the relevant statutory provisions within the field of responsibility of the enforcing authority. The twelve paragraphs

[12] Parl. Deb. H.C. (Written Answers), November 4, 1976. Col. 663.
[13] [1978] I.C.R. 1015.
[14] See "Tomorrow's Inspector—Judge, Jury and Executioner," V.A. Broadhurst in *Process Engineering,* September 1974, pp. 116–119.

(*a*) to (*l*) setting out the various powers are complemented by a "sweeping-up" paragraph (*m*). All of these powers represent a drastic interference with private rights and show that the right to conduct an undertaking without governmental interference is as much part of legal mythology as the belief that "an Englishman's home is his castle." They have not given rise to the same volume of litigation as have the inspections carried out under the American Occupational Health and Safety Act 1970, which has had to be operated against the background of a constitutional "bill of rights."[15] Although we in Britain have nothing comparable to a bill of rights in our domestic law (leaving on one side the European Convention on Human Rights), the powers conferred by section 20 are strictly construed, firstly, because they represent an encroachment on property and freedom and, secondly, because they may become the basis of a criminal prosecution.

The powers in section 20(2) are as follows:

(*a*) at any reasonable time (or, in a situation which in his opinion is or may be dangerous, at any time) to enter any premises[16] which he has reason to believe it is necessary for him to enter for the purpose mentioned in subsection (1) above;

(*b*) to take with him a constable if he has reasonable cause to apprehend any serious obstruction in the execution of his duty;

(*c*) without prejudice to the preceding paragraph, on entering any premises by virtue of paragraph (*a*) above to take with him—

(i) any other person duly authorised by his (the inspector's) enforcing authority; and

(ii) any equipment or materials required for any purpose for which the power of entry is being exercised;

(*d*) to make such examination and investigation as may in any circumstances be necessary for the purpose mentioned in subsection (1) above;

(*e*) as regards any premises which he has power to enter, to direct that those premises or any part of them, or anything therein, shall be left undisturbed (whether generally or in particular respects) for so long as is

[15] The exclusion of federal establishment from inspection is a major limitation in the American system.

[16] This, *pace* Broadhurst (fn. 14, above), includes a private house, but the inspector must be of the opinion that entry is necessary for the purpose of carrying into effect a relevant statutory provision, such as to enforce a duty upon a self-employed person working at home, *e.g.* a homeworker. The term "premises," as defined by s.53(1), includes "any place."

reasonably necessary for the purpose of any examination or investigation under paragraph (*d*) above;

(*f*) to take such measurements and photographs and make such recordings as he considers necessary for the purpose of any examination or investigation under paragraph (*d*) above;

(*g*) to take samples of any articles[17] or substances[18] found in any premises which he has power to enter, and of the atmosphere in or in the vicinity of any such premises;

(*h*) in the case of any article or substance found in any premises which he has power to enter, being an article or substance which appears to him to have caused or to be likely to cause danger to health or safety, to cause it to be dismantled or subjected to any process or test (but not so as to damage or destroy it unless this is in the circumstances necessary for the purpose mentioned in subsection (1) above);

(*i*) in the case of any such article or substance as is mentioned in the preceding paragraph, to take possession of it and detail it for so long as is necessary for all or any of the following purposes, namely—

(i) to examine it and do to it anything which he has power to do under that paragraph;

(ii) to ensure that it is not tampered with before his examination of it is completed;

(iii) to ensure that it is available for use as evidence in any proceedings for an offence under any of the relevant statutory provisions or any proceedings relating to a notice under section 21 or 22;

(*j*) to require any person whom he has reasonable cause to believe to be able to give any information relevant to any examination or investigation under paragraph (*d*) above to answer (in the absence of persons other than a person nominated by him to be present and any persons whom the inspector may allow to be present) such questions as the inspector thinks fit to ask and to sign a declaration of the truth of his answers;

(*k*) to require the production of, inspect, and take copies of or of any entry in—

[17] "Article" as such is not defined in the Act, although "article for use at work" is defined by s.53(1) to mean (a) any plant designed for use or operation (whether exclusively or not) by persons at work, and (b) any article designed for use as a component in any such plant. If, as seems to be intended, "article" includes plant, it is difficult to see how one could take a sample of plant, although one could well see plant being dismantled under paragraph (*h*).

[18] A "substance" means any natural or artificial substance, whether in solid or liquid form or in the form of a gas or vapour.

(i) any books or documents which by virtue of any of the relevant statutory provisions are required to be kept; and

(ii) any other books or documents which it is necessary for him to see for the purposes of any examination or investigation under paragraph (*d*) above;

(*l*) to require any person to afford him such facilities and assistance with respect to any matters or things within that person's control or in relation to which that person has responsibilities as are necessary to enable the inspector to exercise any of the powers conferred on him by this section;

(*m*) any other power which is necessary for the purpose mentioned in subsection (1) above.

Section 20(3) enables the Secretary of State to make regulations as to the procedure to be followed in connection with the taking of samples under (*g*) above, including provision as to the way in which samples that have been so taken are to be dealt with. The powers given in (*h*) above are amplified as follows by subsections (4) and (5)—

(4) Where an inspector proposes to exercise the power conferred by subsection (2)(*h*) above in the case of an article or substance found in any premises, he shall, if so requested by a person who at the time is present in and has responsibilities in relation to those premises, cause anything which is to be done by virtue of that power to be done in the presence of that person unless the inspector considers that its being done in that person's presence would be prejudicial to the safety of the State.

(5) Before exercising the power conferred by subsection (2)(*h*) above in the case of any article or substance, an inspector shall consult such persons as appear to him appropriate for the purpose of ascertaining what dangers, if any, there may be in doing anything which he proposes to do under that power.

Subsection (6) qualifies paragraph (*i*) above, and, as will be seen, paragraph (*g*). The power in this latter paragraph (to take samples) differs from paragraph (*i*) in which danger to health or safety is the test.

(6) Where under the power conferred by subsection (2)(*i*) above an inspector takes possession of any article or substance found in any premises, he shall leave there, either with a responsible person or, if that is impracticable, fixed in a conspicuous

position, a notice giving particulars of that article or substance sufficient to identify it and stating that he has taken possession of it under that power; and before taking possession of any such substance under that power an inspector shall, if it is practicable for him to do so, take a sample thereof and give to a responsible person at the premises a portion of the sample marked in a manner sufficient to identify it.

In a Scottish case, *Skinner* v. *John G. McGregor (Contractors) Ltd.* (Sheriff Court),[19] the failure of the inspector to leave a notice or sample under subsection (6) rendered four pieces of asbestos containing crocodilite to be inadmissible in evidence, the court holding that the general power given by (*g*) above was made more specific by (i) and subsection (6), so that it was improper for the inspector to rely on the "general power" in (*g*). The case illustrates the strict construction of drastic powers reinforced by potential criminal liability.

Section 20 concludes with two procedural subsections, the first of which qualifies the extensive power to interrogate given by paragraph (*j*) above.

(7) No answer given by a person in pursuance of a requirement imposed under subsection (2)(*j*) above shall be admissible in evidence against that person or the husband or wife of that person in any proceedings.

(8) Nothing in this section shall be taken to compel the production by any person of a document of which he would on grounds of legal professional privilege be entitled to withhold production on an order for discovery in an action in the High Court or, as the case may be, on an order for the production of documents in an action in the Court of Session.

The Inspector as Interrogator

An inspector for the purpose of carrying into effect any of the relevant statutory provisions and, in particular making examinations and investigations for those purposes, has the important power conferred by section 20(2)(*j*) to require answers to questions put by him to any person whom he has reasonable cause to believe to be able to give information relevant to such examinations and investigations and to require that person to sign a declaration of the truth of his answers. Whilst this power is undoubtedly of utility in assisting the inspector to ascertain the facts in a particular situation, section 20(7), as we have seen, does not allow an answer to a question put under section 20(2)(*j*)

[19] 1977 S.L.T. 83.

to be "admissible in evidence against that person or the husband or wife of that person in any proceedings." Accordingly, if A and B answer questions and sign statements, A's statement could be adduced in evidence against B, and vice versa, but neither statement could be adduced against the person making it. If the inspector wishes to take a statement from someone whom he intends, or may intend, to prosecute, he must use some other power, probably a general power to see that the relevant statutory provisions have not been infringed; but in such a case it should be noted that as a person, other than a police officer, who has powers of investigating and prosecuting offences, he must so far as may be practicable comply with the Judges' Rules.[20]

Under the Rules, an inspector may question a person, whether suspected or not, from whom he thinks information relating to the commission of an offence may be obtained. As soon as the inspector has evidence which affords reasonable grounds for suspecting that a person has committed an offence, he must caution that person or cause him to be cautioned before putting to him any questions or further questions. The well-known formula runs—

> "You are not obliged to say anything unless you wish to do so, but what you say may be put into writing and given in evidence."[21]

The caution must be administered when a person is actually charged with an offence. At both stages, that is to say, before the inspector has reasonable grounds for suspecting that the person questioned has committed an offence and after that stage (when "hostilities" may be said to commence), the accused, whether taken into custody or not, may remain silent. It is true that it is an offence under section 33(1) of the 1974 Act intentionally to obstruct an inspector in the exercise or performance of his powers or duties, but the silence of the accused cannot be treated as obstruction for two reasons. First, a failure to answer a question is not obstruction in the absence of any legal duty to answer and, secondly, the "right to silence" is recognised by law. Two trade unions (EPEA and ASTMS) have urged their members to have a safety representative (in the case of EPEA) or a branch official (in the case of ASTMS) present when a member is being questioned by an inspector so that they may be

[20] Rule 6. No caution is required where a statement and signed declaration is taken from someone (who need not necessarily be an employee) as a possible "witness."

[21] Rule 2. Failure to administer the caution does not necessarily render a statement inadmissible in evidence.

protected from unwittingly incriminating themselves.[22] Doubtless, a failure or refusal to answer a question obstructs the inspector in a factual sense, but following the principle established in *Rice* v. *Connolly*[23] the failure or refusal to answer questions is not obstruction in the legal sense of that term. If an inspector questions A and B in connection with an offence which he reasonably suspects to have been committed by B, the position would appear to be that B can remain silent and even if he does volunteer a statement it is inadmissible under section 20(7) above, whereas it seems that the inspector can "require" A to answer questions and to sign a declaration of the truth of his answers.[24] The term "require" is clearly stronger than the term "request."

The wide-ranging powers in section 20 do not inhabit a legal vacuum since they are reinforced by the criminal sanctions in section 33,[25] which makes it an offence—

— to contravene any requirement imposed under section 20;
— to prevent or obstruct any other person from appearing before an inspector or from answering any question which an inspector may require under section 20(2);
— intentionally to obstruct an inspector in the exercise or performance of his powers or duties;
— knowingly or recklessly to make certain false statements;
— intentionally to make false entries with intent to deceive
— with intent to deceive to forge or use certain documents; and
— falsely to pretend to be an inspector.

Seizure Powers: section 25

An important power, additional to those already described, is conferred upon the inspectorate by section 25, namely, to seize and render harmless an article or substance which the inspector has reasonable cause to believe, in the circumstances in which he finds it, to be a cause of imminent danger of serious personal injury. He must, if practicable to do so, take a sample of an article forming part of a batch of similar articles, or of any substance, and give a portion of the sample (marked so as to

[22] The advice followed the prosecution of an EPEA member—Health and Safety Bulletin No. 75 (March 1982), I.R.S., pp. 6–7, in which the statements were admitted in evidence by the magistrates although not taken in accordance with the Judges' Rules. It lies with the inspector to decide who shall be present—see s.20(*j*) above.

[23] [1966] 2 Q.B. 414.

[24] We are indebted to a lecture given by Mr. J.D. Riley, Barrister, to the 88th. Environmental Health Congress held at Harrogate, September 28—October 1, 1981, and entitled "Legal Problems Arising from the Enforcement of the Health and Safety at Work, etc. Act 1974."

[25] See p. 162, below.

identify it sufficiently) to a responsible person at the premises where the article or substance was found by him. He must also prepare and sign a report giving particulars of the circumstances in which the article or substance was seized and dealt with by him and give a copy of the same to a responsible person at the premises where the article or substance was found by him and to the owner of the article or substance (unless the former person is also the owner). As one might expect in view of the existing powers held by inspectors little use would appear to have been made of the dramatic powers given to inspectors by section 25 although in a proper case they would, no doubt, be used.

Compulsory Insurance

To deal with the problem caused by uninsured employers who were unable to meet claims for personal injury arising out of and in the course of employment in Great Britain, the Employers' Liability (Compulsory) Insurance Act 1969 requires employers carrying on business in Great Britain to insure against employers' liability. The duty to insure and the duty to display certificates of insurance under the Act are criminally enforceable. The implementation of this Act, the aim of which is compensation and not prevention, has been entrusted to HSE inspectors following an agreement under section 13(1)(*b*), of the 1974 Act which provides for HSE to perform on behalf of the Secretary of State the latter's functions under the 1969 Act. The employer's duties under the 1969 Act and regulations made thereunder have been extended to offshore installations, although certain large enterprises have been exempted from those duties, presumably on the basis that they are large enough to have "self-insurance."

ADMINISTRATIVE SANCTIONS

The Robens Committee found that the criminal prosecution and the limited powers of inspectors to give directions were inflexible and uncertain weapons to deal with unsafe or unhealthy conditions. Their recommendation in paragraph 265 of their Report that much greater reliance should be placed on "non-judicial administrative techniques for ensuring compliance with minimum standards of safety and health at work" finds its fulfilment in sections 18–26 inclusive of the Act, thereby illustrating the march to what the American jurist years ago termed "administrative justice."

In addition to the general powers of inspectors given by section 20, sections 21–26 confer sweeping powers upon inspectors to issue improvement notices and prohibition notices to which must be added the power to seize and render harmless causes of imminent danger. As the volume of governmental regulation increases, it is noticeable that increasing use is made of summary powers to relieve pressure on the courts and tribunals. There is, of course, justiciability with regard to these new summary powers but it is justiciability by appeal on the part of the "offender," as opposed to the prosecution in court of offenders by inspectors. There is no doubt that these new summary powers have proved to be popular with the inspectorates if not with their victims.

The table below shows the relationship between prosecutions and the new summary powers in Great Britain for the period January 1978 to March 1980.[26]

	1978	1979	Jan/Mar 1979	Jan/Mar 1980
PROSECUTIONS				
Cases taken	1,820	1,632	342	272
ENFORCEMENT NOTICES ISSUED				
Statutory Notices				
Improvement	12,156	13,498	3,336	3,452
Immediate Prohibition	2,889	3,111	853	714
Deferred Prohibition	529	538	138	134
All	15,574	17,147	4,327	4,300
Crown Notices	12	29	18	8

The preference for summary powers as opposed to prosecution is seen when the 272 prosecutions are compared with the 4,300 enforcement notices issued over the period January–March 1980.

Improvement Notices: section 21

If an inspector is of the opinion that a person is contravening one or more of the relevant statutory provisions or has contravened one or more of those provisions in circumstances

[26] D.E. Gazette, September 1980, vol. 88, No. 9, p. 1008.

that make it likely that the contravention will continue or be repeated, he may serve on him an improvement notice stating that he is of that opinion, giving particulars of the reasons why he is of that opinion, and requiring that person to remedy the contravention or, as the case may be, the matters occasioning it, within such period (ending not earlier than the period within which an appeal against the notice can be brought under section 24) as may be specified in the notice.

The inspector may only issue an improvement notice in respect of a contravention of a relevant statutory provision (or the continuation or repetition of such a contravention) but need not point to any risk to safety or health.[27] Confronted with what he perceives to be a contravention of the law, the inspector in issuing an improvement notice is in a sense turning a blind eye to the contravention presently taking place and is giving the recipient of the notice an opportunity to mend his ways, a time for repentance which may be extended under section 23(5).[28] Section 21 is, therefore, very much more a "lawyer's section" than section 22 which deals with prohibition notices. In issuing an improvement notice, the inspector cannot seek to improve upon the law. Thus, if an improvement notice is served in relation to a building, the inspector cannot require measures to be taken to remedy a contravention which are more onerous than those necessary to secure conformity with the requirements of any building regulations for the time being in force,[29] unless the statutory provision which is being contravened itself imposes specific requirements more onerous than building regulations currently in force.

The notice may either require the recipient to remedy the contravention or the matters occasioning the contravention or it may go further and add to that requirement directions as to the remedial steps which need to be taken (normally added as a schedule to the *pro forma* notice which is used). The first type of notice may be specific (*e.g.* to place a fence on a machine) or it may be unspecific as to the means to be used to remedy the contravention (*e.g.* to raise the temperature of a workroom). The inspector in *Graham* v. *Holmes*[30] had served a notice alleging three contraventions of section 4 of the Act (which deals with control of premises) one of which was the fitting of unapproved

[27] *Davis and Sons* v. *Environmental Health Department of Leeds City Council* [1976] I.R.L.R. 282 (failure to provide "conveniently accessible" sanitary convenience under s.9 of the Offices, Shops and Railway Premises Act 1963).

[28] See *Baldwin and Partners* v. *Baxendale* (1977), unreported, Carlisle I.T., March 16, 1977,, discussed below.

[29] Building Regulations 1976 (as amended)(S.I. 1976 No. 1676)(England and Wales). See *Manders Property Ltd.* v. *Johnson*, 23.7.80 (C.O.I.T HS/216) 190.

[30] 16.1.78, HS/38151/77, Ashford Tribunal.

bolts to the fire exits of an indoor market. The Kent Fire Brigade and the Home Office had recommended that barrel bolts should not be used and the tribunal, whilst treating these recommendations as useful guidance, amended the notice on the ground that the contravention alleged in respect of the bolting of the doors had not been made out.

The second type of notice may (but need not), in the words of section 23, "include directions as to the measures to be taken to remedy any contravention or matter to which the notice relates; and any such directions:

(a) may be framed to any extent by reference to any approved code of practice; and
(b) may be framed so as to afford the persons on whom the notice is served a choice between different ways of remedying the contravention or matter."

A notice which specifies remedial steps to be taken which are unclear or vague is not thereby rendered invalid in that the power of the industrial tribunal to modify a notice includes power to clarify the terms of the remedial steps without affecting the validity of the notice.[31]

Section 21 is based on the contravention of a relevant statutory provision to which the inspector must refer in his notice giving reasons why he is of the opinion that there has been a contravention. The word "contravention" covers a range of possible situations in which the degree of acceptable opinion must necessarily vary. There is little room for a dialectic in the case of an office staircase, open on one side, for which no form of handrail at all is provided (although there might well be argument about whether or not a handrail which is provided is "substantial" as required by section 16, Offices, Shops and Railway Premises Act 1963, see *Haverson & Sons Ltd.* v. *Winger*.[32] On the other hand the test of "reasonable practicability" which constantly occurs in the formulation of the general duties in the 1974 Act gives more room for "opinion." Tribunals have affirmed improvement notices which have been served in relation to contraventions of section 3 of the Factories Act 1961 and section 6 of the Offices, Shops and Railway Premises Act 1963.[33] These sections (similar in wording) require that effective

[31] *Chrysler (UK) Ltd.* v. *McCarthy* [1978] I.C.R. 939.
[32] 10.9.80 H/S 19536/80/F London (South) Tribunal.
[33] See for example *Broad Sawmills* v. *Radcliffe* (1979) (unreported) (sawmill open to the elements on both sides, but temperature had to be maintained at 10°C): also *Tyrematic* v. *Cottle* (1980) (unrep.). An employee required to work in a workplace with an ambient temperature of 49°F instead of the 60°F required by the Factories Act 1961 was held to be constructively dismissed when he left the premises (*Graham Oxley Tools* v. *Firth*, (1979), Employment Appeal Tribunal (unreported)).

provision be made for securing and maintaining a reasonable temperature in workplaces and that where the work does not involve severe physical effort temperatures of 60°F and 60.8°F, respectively, should be attained after the first hour. In at least two cases, however, the tribunals have decided that it is not reasonably practicable to heat the workroom. In the first case, *MacDermott* v. *Booth*,[34] an inspector argued that the temperature was inadequate in a pipefitting and fabrication shop. The building in question was 9½ acres in extent with a roof height of 50 feet. Not surprisingly perhaps, the notice (which would appear to have required not so much heating as a changed micro-climate) was cancelled. Similarly, an appeal was allowed in *Roadline (UK) Ltd.* v. *Mainwaring*,[35] where the employer had spent a considerable sum of money on the installation of a system to alleviate cold conditions, providing a warm rest room and warm clothing.

As we shall see below, the industrial tribunals are there to curb the zeal of the over-zealous inspector. In *British Airways Board* v. *Henderson*,[36] a notice requiring the fitting of guards to electrical sheaves and cables in lift motor rooms was cancelled. Expert witnesses testified that they maintained 30,000 similar installations in the U.K. and not a single accident had been caused by unguarded sheaves in fifty years and it was significant that neither the British Standards Specification nor a Factories Inspectorate document on lifts mention the necessity to guard sheaves. The requirement of a legal contravention may also be a trap for the inspector as in *G. Dicker and Sons* v. *Hilton*[37] in which a notice issued in relation to an air receiver under section 36 of the Factories Act 1961 was cancelled on two grounds, namely, because the lack of a required certificate did not *per se* render the air receiver dangerous (on which there was no evidence) and also because no employees were involved (thereby excluding the Factories Act 1961 and section 2 of the 1974 Act). It emerged at the hearing that a man working on a car gearbox in the garage was present at the time of the inspector's visit but that he was not an employee of the appellant since he worked as a self-employed person for the appellant and others.

[34] (1977) (unreported), Swansea Industrial Tribunal.
[35] (1977) (unreported), Swansea Industrial Tribunal.
[36] 16.11.78 (C.O.I.T. HS 2/38) and see p.146 below. In *Alfred Preedy and Sons Ltd.* v. *Owens*, 1.5.79 (C.O.I.T. HS/2/102) a notice was cancelled on the evidence of an expert witness that a stone stairway of sound construction but with worn treads was not in a dangerous condition.
[37] 9.2.79 (C.O.I.T. HS/2/102).

Prohibition Notices: section 22

Section 22 confers upon a duly appointed inspector the more drastic power to notify the discontinuance of an activity carried on or about to be carried on by or under the control of any person, being an activity to or in relation to which any of the relevant statutory provisions apply or will, if the activity is so carried on, apply, if as regarding such activity the inspector is of the opinion that it does or will involve a risk of serious personal injury. The inspector who decides to issue such a prohibition notice must state that he is of the said opinion, must specify the matter giving rise to the risk and any contravention of the relevant statutory provisions (where such is the case) and must direct the discontinuance of the activity in question until such time as the matter giving rise to the risk or any contravention specified above is remedied.

Section 22(4) permits a notice to be issued with immediate effect if the inspector is of the opinion, and states it, that the risk of serious personal injury is, or, as the case may be, will be imminent. Otherwise, the notice will come into operation at the end of a period specified in the notice. The issuance of an "immediate prohibition notice" is clearly reserved for extreme and pressing cases, unlike the "deferred prohibition notice" which, like the improvement notice (which cannot be issued with immediate effect), gives time for remedying matters. Unlike an improvement notice, a prohibition notice is linked to the test of "serious personal injury" whether or not accompanied by a contravention of a relevant statutory provision.[38] Another difference is to be found in the ability to use a prohibition notice to deal with a prospective risk, such as a dangerous machine which is being installed but which has not yet been brought into use.

Section 23 supplements sections 21 and 22 by further provisions, some of which are common to both sections. As in the case of the improvement notice, a prohibition notice may give directions as to the measures to be taken and such directions may be framed by reference to any approved code of practice and may afford to the person on whom the notice is served a choice between different ways of remedying the contravention or matter. Before issuing a notice concerning measures to be taken in the case of escape from premises by reason of fire, the inspector is required by subsection (4) to consult with the fire authority. As regards an improvement notice or a deferred prohibition notice, subsection (5) of the section allows the

[38] See below on appeals to tribunals.

inspector to withdraw the notice which has been served at any time within the stipulated period, and to extend or further extend that period at any time when an appeal against the notice is not pending. The inability of the inspector to grant an extension of time once an appeal is pending is explained by the power of the tribunal to modify the notice by extending the time for compliance.

If, as is the case, an inspector cannot withdraw a notice, or extend a time-limit therein, whilst an appeal is pending, it would seem that once the tribunal has affirmed the notice, but with extension of time-limits, the inspector can do neither of these things. Whilst, undoubtedly, an inspector cannot "withdraw" a notice affirmed by a tribunal but not complied with, doubt has been cast by one decision in which the following opinion was given—

> "In connection with these dates (*i.e.* the dates extended by the tribunal), we perhaps ought to mention lastly that on our interpretation of Section 23(5)(b) of the 1974 Act that now that this appeal ceases to be pending, the Factory Inspector still has powers on application to him, to extend or further extend the dates in the Improvement Notice if good cause is shown to him for that. The fact that the matter has been taken to an Industrial Tribunal and that dates have been fixed by that Tribunal does not oust the Inspector's jurisdiction under Section 23(5)(*b*)."[39]

Whilst it is true that an appeal "is not pending" once it has been determined by the tribunal, it is respectfully doubted that an inspector, who cannot extend the time-limit once an appeal is pending, can do so once the tribunal has itself determined the proper time-limit.

An improvement notice relating to a building, or any matter connected with a building, to which any of the relevant statutory provisions apply, cannot direct measures to be taken which are more onerous than those necessary to secure conformity with the requirements of any building regulations for the time being in force to which that building or matter would be required to conform if the relevant building were being newly erected, unless the statutory provision in question imposes specific requirements more onerous than the requirements of any such building regulations to which the building or matter would be required to conform as aforesaid. An example is provided by *Baldwin and Partners* v. *Brazendale* (see fn. 39), in which this provision (section 23(3)) afforded no defence

[39] *Baldwin and Partners* v. *Brazendale* 16.3.77, HS 42194/76.

inasmuch as a factory which was no doubt built in conformity with building regulations was still subject to more onerous requirements imposed by section 1(3) of the Factories Act 1961 concerning the painting, colour washing or whitewashing of factory walls.

Persons on whom notices are served

The service of notices (not confined to those under discussion) is governed by section 46 of the Act which provides that a notice required or authorised to be served on or given to a person other than an inspector may be served or given by delivering it to him, or by leaving it at his proper address, or by sending it by post to him at that address. In the case of a body corporate, the notice may be served on or given to the secretary or clerk of that body; and in the case of a partnership it may be served on or given to a partner or a person having the control or management of the partnership business (or, in Scotland, the firm). It might be added here that whereas an improvement notice must be served on the person contravening a relevant statutory provision, a prohibition notice may be issued for example to a manager of a company or partnership whose directors or partners are not available if the manager carries on or controls the carrying on of the activities in question.

In several of the cases, an appeal has been lodged against a notice on the ground that the appellant is not responsible for premises to which the notice relates, *e.g.* that a notice concerning a leaking roof should not have been served upon an employer because a local authority, from which the premises had been leased, was allegedly liable for dilapidations.[40] Improvement notices necessarily, and prohibition notices possibly, involve contravention of the "relevant statutory provisions" *i.e.* not only provisions in Part I of the 1974 Act (including health and safety regulations) but also the existing statutory provisions consisting of the Acts and subordinate legislation made thereunder as set out in Schedule 1 to the 1974 Act. In other words, the notices are not limited to the new provisions in the 1974 Act but may be served on those who contravene any of the host of provisions in the "old" statutory provisions as listed in Schedule 1. Thus, section 62 of the Offices, Shops and Railway Premises Act 1963 places liability for contraventions relating to premises covered by that Act upon the "occupier," unless the Act makes some other person liable either in addition to or instead of the

[40] *Siveyer* v. *Randall* (1978) unrep., where this argument failed.

occupier.[41] As was indicated in Chapter 1, the codes place greater emphasis upon spatial and functional criteria than does the 1974 Act, with the consequence that the improvement notices, themselves introduced by the 1974 Act in their present form, must respect those spatial and functional criteria.

The 1974 Act itself is not entirely free from limitations of this sort. Section 2(2), as we have seen, refers to "any place of work under the employer's control" somewhat like section 4(2) which refers to "each person who has, to any extent, control of premises" to which that section applies. "Control" is a concept wider in ambit than "occupation" or "ownership."[42] Premises may constitute not only a "place of work" but also a "system of work" such as an oast house used for drying hops.[43]

Prohibition notices are linked to activities carried on or about to be carried on by or under the control of any person which involve risk of serious personal injury. It may be that such a risk involves a contravention of a statutory provision in which event the notice may involve questions relating to the extent of premises, their occupation and control. The great majority of notices of either sort are served on employers. Nevertheless, we have seen in our discussion of section 7 and 8 that employees and those who misuse safety, health or welfare facilities may commit criminal offences. Whilst, it may be true that inspectors prefer to control activities through employers and controllers of premises, they are at liberty to issue notices directly to employees and others in cases falling within section 21 and 22.

APPEALS AGAINST IMPROVEMENT AND PROHIBITION NOTICES

Sections 21 and 22 bestow radical powers on inspectors which when exercised to promote health, safety and welfare at work may have far-reaching consequences not only for those to whom the notices are served but also for others, such as employees, whose interests may be indirectly affected. The sections place a great premium on the judgment and good sense of inspectors who, like the rest of us, are human and therefore subject to error. Section 24 accordingly provides a right of appeal to an

[41] *McNeil* v. *Wane* (1978), unrep., where an appeal was allowed. In *Westergren* v. *Cox* (1978), unrep., the improvement notice was cancelled because it related to premises not covered by the 1963 Act.

[42] In *Morris* v. *Wilkins* (1979), unrep., the tribunal inspected a lease and ascertained that responsibility for a badly leaning wall was shared by the appellant and another person. Duties under the Factories Act 1961 are cast upon the occupier, but see the responsibility of *owners* of tenement factories (s.121 of that Act).

[43] *Bartlett and Son Ltd.* v. *Newble* (1979) unrep., where a prohibition notice on the use of an oast house was affirmed.

industrial tribunal against an improvement notice or a prohibition notice which has been served or against the terms of such a notice, such as the period (if any) after which it is expressed to take effect.

Judging from the Table below, it would seem that the great majority of appeals are withdrawn before hearing, possibly because appeals lodged in hot blood are abandoned when calmer counsels prevail—67 of the 89 appeals for 1980. The chances of being upheld on appeal are slim, although as noted below, tribunals are more willing to modify a notice whilst at the same time affirming it.

Appeals against Enforcement Notices, 1980[44]

Result	HSE Notices (Factory and Agricultural Inspectorates)	Local Authorities	Total
Appeals upheld	1	3	4
Appeals dismissed	1	4	5
Notices upheld with modification	6	7	13
Appeals withdrawn	28	39	67
Total appeals	36	53	89

In the case of an improvement notice, the effect of an appeal is to suspend the operation of the notice until the appeal is finally disposed of or withdrawn; whereas in the case of a prohibition notice, the operation of the notice is suspended if, but only if, on the application of the appellant the tribunal so directs (and then only from the giving of the direction). One or more assessors may be appointed to assist a tribunal in proceedings under section 24 although it would seem that little, if any, use has been made of this facility, possibly because the tribunal as an "industrial jury" already has members with "industrial" experience and also because expert witnesses may lead evidence of a technical nature and thereby give guidance to the tribunal.

The tribunal seised of an appeal may—

(i) cancel the notice; or
(ii) affirm the notice; or
(iii) affirm the notice with such modifications as the tribunal may in the circumstances think fit.

The term "modifications" is defined in section 82(1)(*c*) to include "additions, omissions and amendments. The whole purpose of an

[44] Health & Safety Information Bulletin No. 71, November 1981, p.12.

appeal is to allow the tribunal the opportunity to review the situation. It was for this reason that the Divisional Court in *Chrysler (UK) Ltd.* v. *McCarthy* treated as being out of order an appeal from a tribunal on the preliminary point that an improvement notice was void by reason of imprecision and vagueness, Eveleigh J. saying—

> "What should have happened was that the matter should have been investigated as to the facts. Then, after the tribunal had exercised or failed to exercise its powers under section 24, if there had remained any cause for complaint the Appellants would then be in a position to come to this Court and argue their case."[45]

The tribunal may correct minor technical defects in a notice, or even change the requirements of the safety or other measures specified in the notice, but it cannot modify the notice so as to introduce a duty to comply with some provision in the Act which was not raised in the notice itself.[46]

Before proceeding to see how tribunals have dealt with appeals under a variety of grounds, it is necessary to give the preliminary warning that the doctrine of precedent does not operate, at least formally, with regard to tribunal decisions which in legal theory should be confined to their own peculiar facts. It is noticeable that the tribunals in exercising their separate jurisdiction over employment protection rights have been allowed to arrive at a "range" of decisions on facts with diminished oversight by the Employment Appeal Tribunal of decisions reasonably arrived at within that "range." If, as is here suggested, the concept of a reasonable range of decisions applied to appeals under section 24, there ought to be no surprise that two tribunals on similar facts might arrive at different conclusions within the permissible range of reasonable conclusions, although a conclusion which no tribunal could reasonably arrive at on the facts would be liable to be struck down for error of law. Nevertheless, it will be useful to essay some tentative conclusions from tribunal decisions if only because *stare decisis* is to a large extent a psychological manifestation of a human desire for patterned regularity and order, dealing first with improvement notices.

Appeals against improvement notices

Mandatory standards. Tribunals are understandably reluc-

[45] [1978] I.C.R. 939, at p. 942. There is an appeal for error of law to the Queen's Bench Division of the High Court: see *Walting* v. *William Bird and Son (Contractors) Ltd.* (1976) 11 I.T.R. 70.
[46] *British Airways Board* v. *Henderson* [1979] I.C.R. 77.

tant to disturb improvement notices given for a failure to implement specific statutory standards in which there is little room for argument. In one case, an improvement notice requiring the replacement of a cracked wash-basin by a firm with an excellent record was left undisturbed despite the minor nature of the infringement.[47] Thus, tribunals have refused to interfere in connection with breach of a mandatory regulation in which there is no room for the tribunal to exercise discretion,[48] where a prescribed minimum temperature is required by the Woodworking Machines Regulations (S.I. 1974 No. 903),[49] where the application of these latter regulations seemed odd,[50] where a reasonable temperature was not maintained as had been the case for fifteen years.[51] The farmer who considered that "The Law is for fools to obey and wise men to break" discovered, like Mr. Bumble, the parochial beadle, that a notice based upon a breach of the law could not be disturbed. He had used a rotary grass cutter for many years without incident and objected to a notice requiring a guard to be fixed to the cutter. However, once it was established that the cutter was an agricultural field machine and, as such, required to be guarded, the tribunal and the Divisional Court (to which the farmer acting in person had taken his case) had little choice but to apply the regulation however unreasonable that might seem to the farmer.[52] Once an infringement of the law is seen tribunals are reluctant to be seen as accessories to crime.

Previous record. In line with the preventive aim of the Act an inspector is not obliged to wait until an accident occurs before he condemns a system as unsafe,[53] although the absence of accidents over a period is a matter on which he, and the tribunal to which an appeal is made, can take into account; as occurred in one case in which the time-limit for compliance with the notice was extended in view of the excellent safety record of the company.[54]

Workforce. Protection is needed not only for the prudent, alert and skilled operative who concentrates on his job but also for the careless and inattentive worker whose inadvertent or indolent conduct might expose him to the risk of injury.[55] It does not

[47] *South Surbiton Co-operative Society* v. *Wilcox* [1975] I.R.L.R. 293 (an inspector might be denied costs for a notice *de minimis*).
[48] *Salmon* v. *Cooper,* 30.4.80 (C.O.I.T. H/S 2/199).
[49] *Broad Sawmills Ltd.* v. *Radcliffe,* 27.7.79 (C.O.I.T. HS 2/114).
[50] *Cheston Woodware Ltd.* v. *Coppell,* 26.7.79 (C.O.I.T. HS 2/110).
[51] *Tyrematic Ltd.* v. *Cottle,* 21.2.80 (C.O.I.T. HS 2/185).
[52] *Murray* v. *Gadbury* Q.B.D. (D.C.) HS 13127/77. The farmer had costs to pay for his pertinacity.
[53] *Shipbreaking Q. Ltd.* v. *Tonge,* 17.1.80 (HS 2/162).
[54] *Otterburn Mill Ltd.* v. *Bulman* [1975] I.R.L.R. 223.
[55] *Bottomley* v. *Fellowes,* 9.5.80 (H/S 2/201).

follow, however, that a reduced element of risk because the workforce is intelligent, skilled and well-supervised is a relevant factor to influence a tribunal which is considering whether or not to uphold an improvement notice.[56] In assessing whether or not to affirm an improvement notice the personal views of those employed in the premises are irrelevant,[57] as is the extraordinary fortitude of particular employees inasmuch as an objective assessment must be made from the standpoint of any future and normal employees.[58]

Economic hardship. The cost of complying with an improvement notice may impose a financial burden upon a firm and, in extreme cases, might cause the firm to close down with a consequent loss to those, such as employees, dependent upon that firm. In general, tribunals have been reluctant, at least overtly, to be swayed by economic pleas *ad misericordiam*, at least to the extent of cancelling a notice on this ground, although they are readier to consider extending time-limits. An illustration of this reluctance can be seen in *Harrison (Newcastle-under-Lyme) Ltd.* v. *Ramsey*,[59] in which the tribunal declined to modify an improvement notice relating to the washing down and painting of garage walls, notwithstanding the undoubted financial difficulties which were besetting the firm. It has been seen that the tribunals are reluctant to be *participes criminis* in clear or "open and shut" contraventions of the law such as the breach of the absolute duty under section 1 of the Factories Act 1961 which was involved in the Harrison Case above. In one case, an improvement notice was issued in respect of the overloading of a mill floor, a state of affairs which the company involved blamed on a build-up of stocks caused by the strength of sterling and the low level of activity in the motor-car industry. Notwithstanding the plea that the cost of the remedial work required by the notice would lead to redundancies, the tribunal affirmed the notice because of the safety risk.[60] We shall see that curiously the test of "serious personal injury" used in connection with the graver type of notice—the prohibition notice (discussed below)—allows of a balancing of the risk of injury against the cost of averting that risk. It is more difficult to give relief where the duty is absolute, although here some relief can be given by extending the date for compliance.[61] A tribunal

[56] *Belhaven Brewery Co. Ltd.* v. *McLean* [1975] I.R.L.R. 370.
[57] *Fife Tile Distributors Ltd.* v. *Mitchell* 14.10.77 (HS/6/77).
[58] *Bewlay Properties Ltd.* v. *Jackson*, 7.12.77 (HS/7/77—Scotland).
[59] [1976] I.R.L.R. 135.
[60] *Vinyl Compositions Ltd.* v. *Barnard*, 12.12.80 (C.O.I.T. H/S 3/15).
[61] *Blocking Services Ltd.* v. *North*, 25.10.78 (C.O.I.T. HS 2/30) (duty to fence dangerous parts of machinery absolute—notice affirmed but date for compliance extended by three months).

may take time to consider the cheapest method of compliance with an improvement notice.[62]

Legal defects. An appeal may be successful because of a defect in the inspector's authority or because the notice is not within the terms of the statutory contravention alleged therein. We have seen that some early challenges to the powers of inspectors failed, but in one case, it was determined that an inspector had no authority to issue a notice because his local authority was not an enforcing authority within the Health and Safety (Enforcing Authority) Regulations 1977.[63] If a contravention is based on a statutory provision the notice will be cancelled if it strays outside the terms of the provision, *e.g.* a notice made by reference to section 2(1) of the 1974 Act (which adumbrates the duties of employer *to employees*) will be cancelled if it shown that there were no employees involved at the material times—only independent contractors.[64] If an inspector decides to exercise his discretion under section 20(3) of the Act so as to include direction concerning the remedial measures which need to be taken he must confine himself to requirements which he is entitled in law to impose.[65] In one case, the tribunal cancelled a notice served under section 6(1) of the Offices Shops and Railway Premises Act 1963 and applied the exception in section 6(3)(*b*), which exempts from the reasonable temperature requirement a room in which the maintenance of a reasonable temperature would cause deterioration of goods, provided employees working in it have "conveniently accessible means of enabling them to warm themselves."[66] However meritorious the inspector's emphasis upon health and safety may be he runs the risk of a successful appeal against his notice if no contravention of the law is made out.[67] The risk is less but still there if he insists on too high a standard of compliance,[68] or the tribunal interprets legal requirements in a less draconian manner than the inspector as in one case in which an inspector's notice requiring balustrades to ramps in a multi-storey parking building was restricted to two floors used by shoppers on the ground that the relevant building regulations required a guard against reasonably foreseeable risks and did not require the provision of balustrades at all levels.[69] Lest the impression be gained that

[62] *Brown* v. *Stevenson* 30.3.78 HS/2/78.
[63] *Central Tyre Co. (South Side) Ltd.* v. *Ralph*, 15.9.78 HS 17863/78.
[64] *T.W. Enamellers Ltd.* v. *Chapman* 19.5.77: on s.36 of the Factories Act 1961, see *George Dicker and Sons* v. *Hilton*, 9.2.79 (C.O.I.T. HS/2/102). See above, p. 140.
[65] *T.C. Harrison (Newcastle-under-Lyme) Ltd.* v. *Ramsey* [1976] I.R.L.R. 135.
[66] *NAAFI* v. *Portsmouth City Council*, 8.1.79 (C.O.I.T. HS/2/54).
[67] See *Cadman* v. *Johnson* 29.1.81 (C.O.I.T. H /S3/18): *Lewis* v. *Sweet* 18.2.81 (C.O.I.T. H/S3/19).
[68] *Chethams* v. *Westminster City Council* 13.12.77 (C.O.I.T. H/S1/113).
[69] *Manders Property Ltd.* v. *Johnson* 23.7.80 (C.O.I.T. H/S2/216).

the tribunals apply the law in a mechanistic and wooden fashion, cases show that they have held a notice to be properly served when the person accepting it only worked for the firm on a contract for services,[70] and, as we have seen, they have shown that a notice can be amended without annulling it, as where it schedules unclear or imprecise directions or otherwise requires modification. A tribunal robustly rejected the strained argument in one case that premises of a firm based at Yeadon, near Bradford, were a shipyard and, as such, outside the application of the Factories (Cleanliness of Walls and Ceilings) Order 1960, as amended,—Yeadon-by-the-Sea![70] It would seem that legalism, amounting to a storm-in-a-teacup, can creep in as would seem to have occurred with the service of a notice with great alacrity on a recently-arrived firm relating to the painting of walls, which involved the tribunal in carrying out its own tests on the porosity of the building materials used in the walls in private; after which the notice was affirmed with modification. In an appeal against the decision of the Industrial Tribunal,[71] held in the Queen's Bench Division of the High Court on February 23, 1982, Forbes J., quashing the findings of the tribunal as unsafe, declared, according to local press reports, that it was "wholly objectionable" that the tribunal should experiment privately on the building blocks which had been used. As a consequence, the respondents, Stockton-on-Tees Borough Council, were required to pay the High Court costs, estimated at £2,500, a costly refutation for local ratepayers of the legal maxim *de minimis non curat lex* (the law does not concern itself with trifles).

No estoppel. One would have thought that a person who complies with an inspector's advice would be immune from a notice but this is not necessarily so. A firm which installed a safety system in accordance with the advice of the Fire Officer and HMFI, was required by the latter to install a new safety system. HMFI had changed its mind. The tribunal reluctantly decided to affirm the notice influenced by the overriding need to ensure safety as the paramount objective. This case shows that the tribunals (and courts) are reluctant to raise an estoppel against public officials in the discharge of public duties, notwithstanding the economic loss which may be caused to those who rely on assurances from those officials.

Extension of time. Although a tribunal will not hesitate to

[70] *Booker Wellman Ltd.* v. *Micklethwaite* 28.11.78 (C.O.I.T. HS2/46).
[71] *Datsun Teesside Ltd.* v. *Stockton-on-Tees Borough Council* Case No. HS/25442/80; 28596/80; 28597/80 (C.O.I.T. No. 9/147/36).

overturn the inspector's opinion which it considers to be erroneous or too ambitious,[72] the onus of proof on the appellant is high bearing in mind that tribunals do not have the same familiarity with health and safety appeals as they do with, say, unfair dismissal complaints. It is, perhaps, still too early to make categorical statements about appeals so far heard although it has been noted that most appeals are unsuccessful—a view, however, based on the early and very limited sample taken. The inference has been drawn that the unsuccessful appeals show how well the inspectors are performing. Be that as it may, it does seem that the tribunals are readier to modify a notice, mostly by granting extra time for compliance, than to quash it entirely. For example, a date for compliance was extended in one case for eight months because of financial considerations although it should be added that the safety elements in the notice had been complied with leaving only the welfare requirements outstanding.[73] In another case, additional time was allowed for the completion of work, because the firm in question was a lessee who needed both his landlord's permission and planning permission before a covered walkway could be erected as required by the improvement notice.[74] The tribunal is entitled to look at the history of communications between the inspector and the appellant in that the latter may have had ample time to comply with oral and written advice and warnings before the issue of the improvement notice.

Appeals against prohibition notices

The more drastic step of a prohibition notice is reserved for cases in which the inspector apprehends a risk of serious personal injury. The statistics[75] show that six times as many "immediate prohibition notices" are served as "deferred prohibition notices" showing that cases of imminent risk predominate. The tribunal is not bound to accept without cavil the *ipse dixit* of the inspector as to the risk,[76] and will cancel a notice based only on a contravention for which an improvement should have been used.[77] The inspector should identify the person against whom the notice is intended to take effect and runs the risk that his notice may be cancelled where he issues a fusillade of notices as

[72] See for example *Warner Group of Companies* v. *Smith* 18.10.79 HS2/134 (contractor proved that fencing off a roadway (as inspector had required) would be more dangerous than not doing so).
[73] *R.A. Dyson and Co. Ltd.* v. *Bentley* 26.11.79 (HS 2/151).
[74] *Porthole Ltd.* v. *Brown* 7.3.80 (C.O.I.T. H/S 2/186).
[75] See p. 145, above.
[76] *Brewer* v. *Dunstan* 18.0.78 (C.O.I.T. HS 2/20).
[77] *Russell* v. *Kelly* 15.6.77 HS/2/77 (Scotland).

happened in one case in which the inspector had served notices on all five firms involved in a building operation following the fall of a roof truss which was being lifted by a crane. Notices involving sections 2(1), 3(1) and (2) had been served on all five firms involved in the work but the tribunal disapproved rightly of this "broadside" approach, pointing out that the appellants the crane hire people, were not the proper people to comply with a requirement as to a safe system affecting the work as a whole which was the responsibility of the person controlling the activities.[78] The tribunal in that case noted that there was no provision in the statute whereby persons on whom notices have been served will be informed when the notice has in fact lapsed. The statute is similarly silent as to the ways in which notices generally cease to have effect (the same applying to improvement notices). There is no provision for some certificate of compliance by the inspector to show that the steps taken meet with his approval. An illustration of the complications which can arise is provided by the case of *Box* v. *Ware*[79] in which an appeal was laid against a Chief Environmental Health Officer who had issued a prohibition notice concerning emissions of toxic gases from two gas-fired boilers. B lost no time in putting in his appeal against the notice and in putting matters right. The tribunal determined that the issue of the notice was justified and went on the criticise the lack of communication between the Chief Environmental Health Officer and the appellant, saying that—"Once the respondent knew, or should have known, that matters were put right. . . . there should have been some communication in the form of a letter to the appellant to say so," and speculating that, "It may be that, as a result of this case, there will be formulated some form of streamlining their means of communication so that in a case of this nature, the authorities should themselves be sure to advise the appellant that all necessary steps had been taken." No order for costs was made.

The inspector may withdraw a prohibition notice (and an improvement notice) which is not to take immediate effect provided he does so before the end of the period specified therein (including any extended period).[80] An inspector might theoretically (and unrealistically) withdraw a notice and substitute a different notice, a form of vaccillation which would cause resentment and possibly loss, where an employer had complied with the first notice before it was withdrawn. If an appeal were

[78] *Stephensons (Crane Hire) Ltd.* v. *Gordon* (H.M. Inspector of Health and Safety) 17.8.77 HS/8868/77.
[79] 7.5.81 HS 7446/81.
[80] s.23(5).

to be lodged against the second notice, the inspector would be likely to receive scant sympathy from the tribunal and it is possible that he (and HSC or HSE) might find themselves subjected to an investigation by the Parliamentary Commissioner for Administration.[81]

Onus on appellant. The appellant has a heavy onus to discharge if he wishes to rebut the prima facie case of risk of serious personal injury raised by the inspector. As in the case of improvement notices, an accident-free history is not enough,[82] or that the appellant had done everything in his power to meet the requirements of the inspectorate but the risk of serious personal injury remains.[83] *A fortiori*, the appellant is likely to command scant sympathy, as where a farmer is shown to have missed three opportunities to carry out remedial work,[84] or an enterprise allows of the emission into the atmosphere of red lead without much thought or attention to the matter except when it is too late.[85]

Espionage rather than inspection. Although the Act is silent on the matter, the tribunals have shown antipathy to the gathering of information by an inspector in a furtive manner as occurred in one case where the inspector noticed that passengers were being transported on the footplate of a steam train.[86] The notice in that case was cancelled, not because of the clandestine manner in which the information had been garnered, but because the inspector could not adduce sufficient evidence as to the risk of serious personal injury.

Economic hardship. Once a risk of serious personal injury is shown, the tribunals show great reluctance to allow economic factors,[87] such as loss of production and the wages of workers,[88] to defeat a notice. Nevertheless, either overtly or covertly, the economic factor must come into the equation unless the remotest and most contingent of health and safety risks must be removed by measures involving great cost and social disruption. The risk may of course be small, as in *Nico Manufacturing Co. Ltd.* v.

[81] See on the Ombudsman, Schedule 1 to Parliamentary Commissioner Act 1967 as amended. At the end of 1981 only four cases were under investigation for maladministration alleged against HSE—Annual Report of Parliamentary Commissioner for Administration for 1981, H.M.S.O., 1982. EHO's are subject to the Local Ombudsmen: see Local Government Act 1974, Part III.

[82] *Hoover* v. *Mallon* 8.12.77 (C.O.I.T. 696/225).

[83] *George Dorset Ltd.* v. *Hill* 6.11.78 (C.O.I.T. H/S 2/44).

[84] *C.R. Bartlett* v. *Newble* 4.7.79 (C.O.I.T. H/S 2/107).

[85] *Bourne Chemical Industries Ltd.* v. *Affleck* 9.11.79 (H/S 2/147).

[86] *Bressingham Steam Preservation Co. Ltd.* v. *Sincock* 5.4.77 (HS 27434/76)—costs awarded against inspector on County Court Scale 4 which included costs of two counsel.

[87] *Wilkinson* v. *Fronks* 29.6.78 (C.O.I.T. HS 1/241).

[88] *Hoover* v. *Mallon* 8.12.77 (C.O.I.T. 696/225).

Hendry,[89] but the occurrence of the risk (disintegration of a power press) would involve risk of serious injury. Any assessment of risk involves balancing different considerations—the likelihood of the risk occurring, the likely consequences of that occurrence, the cost, time and effort in attempting to remove or minimise the risk and so forth. That the tribunals can take into account wider factors is shown in *Otterburn Mill Ltd.* v. *Bulman*,[90] in which a prohibition notice was relaxed for a time in relation to one carding machine so as to avoid "serious embarrassment" to the respondents.

No estoppel. Harsh as it might appear to the recipient of a notice, the cases lean against raising an estoppel against an enforcing authority which can demonstrate risk of serious personal injury. In *Hixon* v. *Whitehead*,[91] H set up business in reliance so far as safety matters were concerned on a letter from an Environmental Health Officer permitting him to store 4,000 kgs of LPG. After complaints by residents, the local Council's health and safety inspector issued a prohibition notice limiting storage to 500 kgs and this notice was upheld by the tribunal on the paramount need to avoid serious injuries.

If appeal to the industrial tribunal is the remedy open to a person upon whom a notice has been served, prosecution for non-compliance is the course open to the enforcing authority in cases where notices are disregarded in whole or in part. The integrity of the system of administrative sanctions would be undermined if cases of non-compliance were passed over in silence. In one case, the tribunal whilst recognising that discretion to prosecute lay with the inspector, suggested that prosecution would not come amiss in the circumstances of that case if the notice were not to be implemented within the time required.[92] In some cases, such as the occurrence of a serious accident, both prosecution and a notice might be used, although it is clearly desirable that any appeal against the latter should be determined, or the period for appeal allowed to run out, before criminal proceedings are instituted.

Appeals procedures

A person on whom an enforcement notice has been served may within 21 days from the date of the service appeal to an industrial tribunal. Industrial Tribunals were created by the

[89] [1975] I.R.L.R. 224.
[90] [1975] I.R.L.R. 223.
[91] 22.7.80 (C.O.I.T. H/S 2/218): IDS Brief No. 190, 12–13.
[92] *Campion* v. *Hughes* [1975] I.R.L.R. 291.

Industrial Training Act 1964 and their jurisdiction extends over several different fields. There are two separate systems of tribunals; one for England and Wales with its head office in London, and one for Scotland, with its head office in Glasgow. A tribunal generally consists of a legally qualified chairman and two members. The two lay members are chosen from a panel of lay persons who are appointed to it by the Secretary of State for Employment on the basis of their knowledge and experience of employment in industry. A person with special knowledge or experience of the subject matter of the appeal may be appointed by the President or a nominated chairman to sit with the Tribunal as an assessor.[93]

Separate rules of procedure have been made for England and Wales and for Scotland in order that appeals against improvement notices and prohibition notices might be determined. These are The Industrial Tribunals (Improvement and Prohibition Notices Appeals) Regulations 1974 (S.I. 1974 No. 1925) and The Industrial Tribunals (Improvement and Prohibition Notices Appeals)(Scotland) Regulations 1974 (S.I. 1974 No. 1926).

An appeal must be initiated by any appellant sending to the relevant Secretary of Tribunals a notice of appeal in writing setting out the name of the appellant and his address for the service of documents; the date of the improvement notice or prohibition notice appealed against and the premises or place concerned; the name and address of the respondent and the particulars of the requirements or directions appealed against. So that the inspector may know the case which he has to meet, the notice of appeal should also indicate the grounds of appeal. If fresh grounds of appeal are raised at the hearing, the tribunal is likely to adjourn the proceedings to afford the Inspector an opportunity to consider and respond to the new points.

The notice of appeal must be sent to the Secretary of the Tribunals within 21 days of the date of the service on the appellant of the notice appealed against.[94] An appellant may apply for an extension of time either before or after the expiry of the twenty-one day period if it is or was not reasonably practicable[95] for an appeal to be brought within that time.

When the Secretary of the Tribunals receives a notice of an appeal against an improvement or prohibition notice, or, in addition to the latter, an application is made to the tribunals for the notice to be suspended,[96] he must enter particulars of it

[93] So far as is known, the power to appoint an assessor has not so far been exercised.
[94] The time limit runs from the date of the receipt of the notice: *D.H. Tools Co.* v. *Myers* (1978) 1 D.S. Brief 138.
[95] Extensions of time are only rarely granted for cogent reasons.
[96] See section 24(3)(6) of the Health and Safety at Work etc. Act. 1974.

against the entry in the Register and send a copy to the respondent. The President or nominated chairman will then fix a date, time and place for the hearing of the appeal and the Secretary of the Tribunals will notify each party accordingly.

A tribunal has wide powers to demand the attendance of witnesses and the production of documents. A tribunal may on the application of a party either by notice to the Secretary of Tribunals, or at the hearing, require a party to provide for the benefit of another party further particulars of the grounds on which he relies and of any facts and relevant contentions; grant to a party such discovery or inspection of documents as might be granted by a county court; and require the attendance of any person as a witness or require the production of any document relating to the matter to be determined, and may appoint the time or place where any of the above acts shall be done.

A tribunal cannot order the production of a document certified by the Secretary of State as one whose production would prejudice the interests of national security.

The rules permit hearings to be in private either on grounds of national security or to avoid serious prejudice to the interests of the undertaking.

If either party wishes to make representations in writing to the tribunal these must be sent to the Secretary of the Tribunals and the other party at least seven days before the hearing. A party may appear before the Tribunal in person or may be represented by counsel or by a solicitor or any other person whom he desires to represent him.

At the hearing each party may make an opening statement, give evidence, call witnesses, cross examine witnesses and address the tribunal. The Tribunal may require witnesses to give evidence on oath or affirmation.[97] Tribunal decisions, which are recorded in a document signed by the chairman, may be taken by a majority and if the tribunal has only two members the chairman will have a second or casting vote.[98]

The making of an appeal suspends an improvement notice[99] until the appeal is finally determined or, if the appeal is withdrawn, until the withdrawal of the appeal. It does not, however, automatically postpone the date given for the completion of the work. Normally if the date for completion has passed by the time the appeal is determined or it is then impossible for the work to be completed in time the tribunal will award a modified date. Thus, in *Campion* v. *Hughes*,[1] the tribunal, whilst

[97] Tribunals have on a number of occasions visited premises to assess matters. See for example *Wilkinson* v. *Fronks* (1978) IDS Brief 143.
[98] Costs may be awarded to the successful party under Rule 13.
[99] See section 24(3)(*b*) of the Health and Safety at Work etc. Act 1974.
[1] [1975] I.R.L.R. 291.

unsympathetic to the respondents because of delay, allowed the time-limit in an improvement notice to be extended so that the consent of a borough council could be obtained to the provision of a fire escape leading to its land. In the case of a prohibition notice,[2] the bringing of an appeal does not suspend the effect of the notice unless the tribunal so directs and then only from the giving of the direction.

Either party may apply to the tribunal for a review of its findings within fourteen days of the date of the entry of a decision in the Register. The application must be in writing and must state the grounds for review in full.[3] Relevant grounds include

(a) that the decision was wrongly made as a result of an error on the part of the tribunal staff; or
(b) that a party did not receive notice of the proceedings leading to the decision; or
(c) that the decision was made in the absence of a party; or
(d) that new evidence has become available since the making of the decision provided that its existence could not have been reasonably known of or foreseen; or
(e) that the interests of justice require such a review.

Appeals may be made from a tribunal on a point of law only. The appeal is to the High Court[4] or Court of Session and must be made within forty two days of the date of the entry of the decision in the Register.

Power to indemnify inspectors

It is possible that an inspector might incur liability in respect of an act or omission in the execution or purported execution of any of the relevant statutory provisions. Thus, the inspector might be liable for that most fertile of torts, the tort of negligence. Leaving aside the possible liability of the inspector to his employing authority,[5] the liability will be incurred to those to whom he owes a duty of care and whose harm has resulted from the breach of that duty; the test of "foreseeability" being used both to delimit the scope of the duty and the remoteness of harm. In such an event, section 26 grants a discretion to his enforcing authority to indemnify him against the whole or part of any damages and costs or expenses which he

[2] See section 24(3)(*b*), *ibid.*
[3] See Rule 12 of the Regulations, *op. cit.*
[4] Tribunals and Inquiries Act 1971; R.S.C., Orders 55 and 94.
[5] *Lister* v. *Romford Ice and Cold Storage Co. Ltd.* [1957] 1 All E.R. 125, unless the inspector can be likened to the holder of an office in the execution of which he does not act as a servant.

may have been ordered to pay or may have incurred, but this discretion is subject to two qualifications, *viz.*:

(i) the circumstances are such that he is not legally entitled to require his enforcing authority to indemnify him; and
(ii) in any case, the enforcing authority must be satisfied that he honestly believed that the act complained of was within his powers and that his duty as an Inspector required or entitled him to do it.

So far as the general common law is concerned, an action may be brought against a tortfeasor regardless of whether or not he was employed by another. The latter may be liable as a joint tortfeasor (although not directly involved in the act or omission giving rise to the tort) and in most cases it is more profitable to sue the employer than the employee. In the case of the Crown, the constitutional maxim that "The King can do no wrong" precluded Crown servants from falling back on the defence of superior orders (*respondeat superior*) when they themselves were sued in tort, although, since the Crown Proceedings Act 1947, there had been a defined right to sue the Crown in tort. It has, on occasions, been a matter of some difficulty to determine whether or not a particular public corporation or regulatory commission is sufficiently close to the Crown to shelter under "the umbrella of the Crown," but this difficulty is avoided by section 10(7), which expressly states that the functions of HSC and HSE, and their officers and servants, are performed "on behalf of the Crown."

It may seem harsh that a servant, including a public servant, should be exposed to liability by reason of his conduct in the performance of his duties. Some solace, however, may be derived from the implied contractual right of a servant or agent at common law to be indemnified against damages, costs and expenses incurred in the performance of his duties as servant or agent.[6] In one case, it was held that a consulting engineer, sued for libel in connection with matters contained in his report, was entitled to indemnity because he came "within the well-settled rule that an agent has a right against his principal, founded upon an implied contract, to be indemnified against all losses and liabilities, and to be reimbursed all expenses incurred by him in the execution of his authority."[7]

Indemnity under section 26 is only available in circumstances in which the inspector is not legally entitled to require the enforcing authority to indemnify him. It may be that an

[6] *Adamson* v. *Jarvis* (1827) 4 Bing. 66: *Dugdale* v. *Lovering* (1875) L.R. 10 C.P. 196.
[7] *per* Lord Cozens-Hardy M.R. in Re *Famatina Development Corporation Ltd.* [1974] 2 Ch. 271.

inspector is entitled to indemnity by reason of some express provision in his contract or by reason of the implied indemnity described above. The traditional view that a Crown servant does not have a contract with the Crown, so that express or implied terms cannot bind the Crown, has been criticised for stating as a substantive rule, rules which are relevant to enforcement.

If the enforcing authority is a local authority, section 265 of the Public Health Act 1875, as amended by section 27 and 39 of the Local Government (Miscellaneous Provisions) Act 1976, gives protection against personal liability to members and officers for things done bona fide (and without negligence)[8] for the purpose of executing any public general or local Act; a protection which covers Environmental Health Officers in the discharge of their health and safety at work functions.

The discretion given to the enforcing authority enables it to refuse indemnity where an inspector has acted fraudulently, dishonestly or in breach of the criminal law, although it would seem to extend to an inspector who has incurred liability in respect of some act or omission which might not entitle him to indemnity at common law provided that the authority is satisfied as to his honest belief in his powers and duty.

One comparatively recent development may be noted here. This relates to the manner in which public authorities, central and local, have been held liable for the tort of negligence committed in the discharge of their public duties and powers. Starting with the case of *Home Office* v. *Dorset Yacht Co. Ltd.*[9] (in which the Home Office was held liable for damage caused by the escape of Borstal boys) the private tort of negligence has begun to have a public aspect and has been applied to local authorities whose surveyor negligently carried out inspections of building foundations the condition of which later caused economic loss.[10] It mattered not that the builder performed his work carelessly; this circumstance does not absolve the authority from liability. The implications of this development are considerable, quite apart from the extension of the duty of care to include realty as opposed to chattels, and the inclusion of economic loss as a proper head of compensation. The Courts have rejected the argument (based upon a distinction between a "duty" and a "power") that because an authority empowered to act has no duty to act, therefore there can be no liability if it does decide to act and does so carelessly. Lord Wilberforce examined the Public Health Act 1936, in *Anns* v. *Merton L.B.C.*, saying—

[8] See *Bullard* v. *Croydon Hospital Group Management Committee* [1953] 1 Q.B. 511.
[9] [1970] A.C. 1004.
[10] *Dutton* v. *Bognor Regis U.D.C.* [1972] 12 Q.B. 373.

> "Undoubtedly, it lays out a wide area of policy. It is for the local authority, a public and elected body, to decide upon the scale of resources which it can make available in order to carry out its functions. . . . How many inspectors, with what expert qualifications, it should recruit, how often inspections are to be made, what tests are to be carried out, must be for its decisions. It is no accident that the Act is drafted in terms of functions and powers rather than in terms of positive duty."[11]

Nevertheless, his Lordship considered to be "too crude" the argument that a public authority could escape liability by simply not inspecting at all.

Lord Wilberforce went on to distinguish between the policy aspects and operational aspects, of a power exercised under a statute.[12] Policy aspects such as the scale of resources available to carry out powers, the number of inspectors to be appointed and the type of inspections to be made were primarily matters for the enforcing authority and not for the courts although these decisions must be made responsibly. It might be said at this point that such decisions made by local authorities are subject to review under the Act.[13]

Operational aspects such as, for example, did the inspector carelessly perform the task allotted, or, did he fail to inspect in any case or class of case where the authority had declared that it was its policy to do so, are reviewable by the Courts and in such cases the inspector, and in consequence his employer authority, may be liable in negligence.

[11] [1977] 2 W.L.R. 1024, at p. 1034.
[12] S.C., at page 1034, *et seq.*
[13] See section 45, discussed above, p. 000.

Chapter 6

CRIMINAL SANCTIONS

From an early date it was recognised that factory legislation should operate through the criminal law by imposing duties, breach of which was an offence. Our early factory legislation concerned the employment of young persons in mines and textile mills and concentrated on the hours of work. It was later that attention was turned to potential dangers such as the use of unguarded machinery.[1] Because an unenforced law may be disregarded with impunity, provision was made in the first Act of 1802[2] for the appointment of factory visitors, but none were appointed. The genesis of our modern inspectorate is to be found in the Children and Young Persons Act 1833,[3] which provided for the appointment of four inspectors of factories with wide powers and duties. One of these powers was the power to prosecute mill owners who broke the new law and it is interesting to note that parents themselves could also be prosecuted.

It became apparent in the years following the second world war that too much reliance upon criminal sanctions could be harmful to the objective of reducing accidents and disease at work. The criminal law with its threats of fines and imprisonment (the latter a somewhat remote possibility in the present context)[3a] is *par excellence* the external agent and, as such, the antithesis of "self-regulation." The criminal law is essentially concerned with human acts or omissions. Most accidents are caused by inattention or carelessness, rather than by deliberate or reckless wrongdoing for which the criminal law is best suited. Thus, the penalties in the 1833 Act were scaled to the degree of "wilful or grossly negligent" wrongdoing by the factory occupier. The criminal law has always been concerned with the mental state of the person accused of crime—with his *mens rea*. Crimes may be classified according to this mental state. Thus, criminal liability attaches for a deliberate and wilful breach of a statutory duty, *e.g.* the employer who knowingly installs unsafe

[1] Factory Act 1844 (7 & 8 Vict., c.15).
[2] 42 Geo. 3, c.73.
[3] 3 & 4 Will. 4, c.103.
[3a] See, however, *De Charette* v. *Huiles-Goudrono-Dérivés* tribunal de grande instance de Béthune, *Le Monde*, octobre 3, 1975, in which a French employer was imprisoned for his failure to ensure a safe system of work—an event which caused something of a furore, not without ideological overtones.

plant. It may also attach where recklessness, falling short of a
intentional or deliberate mental state, is present (see th
discussion at p. 104). Third, there may be a criminal breach o
duty where has been negligence, *i.e.* a failure to observe th
standard of reasonable care. Fourth and last, criminal liabilit
may be imposed for what is purely inadvertent conduct, that i
to say, conduct which is not deliberate, reckless or eve
negligent—the crime of "strict liability" or "strict prohibition
as it is termed. Whilst, few if any would dissent from attachin
the stigma of criminality to intentional, or even reckless
behaviour, experience tends to show that using the criminal lav
for negligent conduct, or, *a fortiori*, conduct which is neithe
negligent nor deliberate or reckless, is counter-productive.

There is also the consideration that the inspectors who enforc
the law have a dual function, namely, to advise and persuade o
the one hand and to prosecute on the other; too great a relianc
upon the latter makes the former role difficult to achieve
Besides, the laborious business of bringing prosecutions and th
reluctance of magistrates to impose anything like the maximum
possible fines are disincentives towards too free a use of th
criminal law. The Robens Committee declared a break with th
old tradition when it said that the idea of rigorously enforcin
standards through extensive use of legal sanctions "is one tha
runs counter to our general philosophy."[4] The recommendation
of the Committee on the future use of the criminal law now fin
their fulfilment in the Act. These recommendations were, first
that prosecutions should be reserved for offences of a flagrant
wilful or reckless nature. Second, maximum fines should b
increased, with provision for higher penalties for repeate
offences and, third, the directors, managers and operative
responsible for breaches should be made liable to prosecution i
addition to a corporate body which is itself answerable in law.

Offences[5]

Section 33, which creates the offences under the Act, specifies
fifteen categories of offence. Some of these are major substantive
offences, such as failure to discharge one or more of the general
duties in sections 2–7, or to contravene sections 8 or 9, or any
health or safety regulation made under the Act (including a
requirement or prohibition in such regulation). Contravention
of any requirement or prohibition imposed by an improvement
notice or prohibition notice is also made an offence, thereby
showing that the new administrative sanctions conferred upon

[4] Report, para. 255.
[5] See "Criminals at Work," Diana Kloss, [1978] Crim. L.R. pp. 280–286.

inspectors are ultimately dependent upon enforcement through the criminal law (discussed below).[6]

In common with criminal offences generally, the offences under the Act differ according to their gravity. The most serious offences are indictable, that is, triable by judge and jury in the Crown Court which has wide powers of punishment. Following the Criminal Law Act 1977, sections 14, 64(1), offences which are not "triable only summarily" are labelled as "offences triable either way," *i.e.* either before the magistrates' court, or the Crown Court on indictment. These offences were previously known as "hybrid offences" since they might be tried summarily or on indictment. It is provided in the Act that offences specified in sections 33(1)(*e*)(part), (*f*), (*h*) and (*n*) shall be triable summarily only and that offences specified in sections 33(1)(*a*), (*b*), (*c*), (*e*)(part), (*g*), (*i*), (*j*), (*k*), (*l*), (m) and (*o*) are indictable offences.

Section 33 is framed in flexible terms so that most of the offences in that section render the offender liable—

(a) on summary conviction to a fine not exceeding £1,000,[7] or
(b) on conviction on indictment—
  (i) to imprisonment not exceeding two years, or a fine, or both, for the offence specified in section 33(4);
  (ii) to a fine.

Because these offences are not "triable only summarily," they are "offences triable either way" within the meaning of the Criminal Law Act noted above. The definition of "indictable offence" in section 64(1) of that Act refers to an offence which, if committed by an adult, is triable on indictment, whether it is exclusively so triable, or triable either way, from which it will be seen, that technically speaking, most HSW offences are "indictable offences." The procedure for determining the mode of trial of offences triable either way is set out in sections 19–26 of the Criminal Law Act 1977 which, *inter alia*, allow the accused the right to make representations as to the mode of trial and to require trial on indictment if he so wishes (s.21(3)(*b*)). The more serious offences referred to above (see (b)(i) above) carry with them imprisonment as a possible punishment and are set out in section 33(4), namely, contravening relevant statutory provisions by doing without a licence something for which a licence is required, contravening a term, condition or restriction in such a

[6] See p. 164. Indeed, "the issue of at least prima facie criminal responsibility will often be decided by tribunals rather than criminal courts," *per* Diana Kloss *op. cit.* p.282.
[7] The fine of £400 in s.33(3) was increased to £1,000 from July 17, 1978. This subsection covers not only offences under the 1974 Act but offences laid down in the existing statutory provisions. Section 33(3) is subject to s.15(6)(*d*) (on which see footnote 9 on page 164).

licence, offences in relation to explosive articles or substances and the offence of contravening a requirement or prohibition imposed by a prohibition notice (*note,* not an improvement notice). If the offence is tried on indictment, the fine is unlimited. If what is termed the "hybrid offence" is tried summarily, the maximum fine, as has been seen, has been increased from £400 to £1,000.

Health and Safety Regulations made under the 1974 Act may exclude proceedings on indictment in relation to offences consisting of a contravention of a requirement or prohibition imposed by or under any of the existing statutory provisions, the general duties in sections 2 to 9 of the 1974 Act, or health and safety regulations.[8] They may also restrict the punishments which can be imposed for any of the foregoing offences[9] (excluding the maximum fine imposable for conviction on indictment).[10] So far as the less serious offences are concerned section 33(2) specifies offences triable summarily for which the maximum fine has also been increased from £400 to £1000.[11] The continuing offences in section 33(5)(continuing failure to comply with prohibition or improvement notice or court order) are summary offences only and the maximum *per diem* fine has been increased from £50 to £100.[12]

The decision to prosecute

In deciding to prosecute, the enforcing authorities have choices to make. First, it should be recalled that the offences for which they may prosecute are not merely the breaches of duties contained in the 1974 Act but also breaches of any of the existing statutory provisions pending their replacement by new regulations (a hope likely to be deferred so far as wholesale replacement is concerned). As will be seen from the Table below, the more detailed provisions in the codes have led to more informations than the 1974 Act.[13]

Although the number of informations may seem large, it is not the policy of the Executive to prosecute for every breach of health and safety law which comes to its notice, any more than it is the policy of the police to prosecute every motor offence of which they have cognisance. The criteria which fortify a decision to prosecute are (1) deliberate flouting of the statutory

[8] Section 15(6)(*c*).
[9] Section 15(6)(*d*).
[10] Criminal Law Act, Sched. 12.
[11] Criminal Law Act 1977, Sched. 6.
[12] Criminal Law Act 1977, Sched. 1.
[13] Written Answers, Parl. Deb. H.C., June 12, 1978, cols. 363–364.

provisions, (2) recklessness of employers and others in exposing people to hazards and (3) a record of repeated infringements by the person concerned. Naturally, the Executive is particularly concerned with those cases in which death or serious injury has resulted from non-compliance with the law or in which persons have been exposed to serious hazards notwithstanding that no actual injury has resulted.[14]

	1975		1976		1977	
	No. of informa- tions	Average fine £	No. of informa- tions	Average fine £	No. of informa- tions	Average fine £
Factories Act 1961, Offices, Shops & Railway Premises Act 1963, and Regulations thereunder	2,832	70	1,814	81	2,190	92
Health and Safety at Work Act 1974	124	88	364	117	649	107

In deciding between summary prosecution or prosecution for the majority of offences which, as we have seen, are triable either way, the factors affecting the decision are (1) the gravity of the offence, (2) the adequacy of the powers held by the summary court, and (3) the record of the accused and his responsiveness to advice. It will be seen that one criterion which is not used is the seriousness of an injury since this may bear little relationship to the gravity of the legal contravention, if any, or the relative culpability of the offender.[15] The only information[16] concerning prosecutions broken down section-by-section relates to prosecutions taken by the Factory Inspectorate, of which the details (dealing with informations rather than cases (which might cover more than one information)) are as overleaf—

It will be seen that informations laid under section 2 outnumber the total number of informations laid under sections 3–6.

As one might expect, the number of prosecutions on indictment (only eighteen in the period January 1, 1975 to January 1, 1978) represents only a small proportion of the total number of

[14] See HSC Report 1974–1976, H.M.S.O. (1977), Annex 4, p. 26.

[15] Written Answer (Mr. John Grant), H.C. Parl. Deb., July 14, 1976, col. 160. The Court is guided by the provisions of the Criminal Law Act 1977.

[16] Written Answers, Parl. Deb., H.C., February 19, 1982, col. 251.

Prosecutions taken by HM Factory Inspectorate under the Health and Safety at Work Act 1974

Year and section of Act	Informations No.	Informations per cent.	Employers Results Convic-tions	With-drawn	Dis-missed
1975					
2	54	98	52	1	1
3	6	100	4	1	1
4	4	100	4	—	—
6	3	100	3	—	—
7	—	—	—	—	—
1976					
2	184	100	164	8	12
3	36	100	32	1	3
4	9	100	4	1	4
6	11	100	11	—	—
7	—	—	—	—	—
1977					
2	344	99	303	25	16
3	45	100	39	3	3
4	24	100	15	8	1
6	26	100	22	—	4
7	—	—	—	—	—
1978					
2	326	99	280	23	23
3	66	100	58	2	6
4	19	100	17	1	1
6	29	94	20	3	6
7	—	—	—	—	—
1979					
2	263	99	221	22	20
3	64	100	51	6	7
4	23	92	21	—	2
6	29	100	23	5	1
7	—	—	—	—	—
1980					
2	261	99	232	15	14
3	71	99	57	6	8
4	39	100	34	2	3
6	27	100	24	—	3
7	—	—	—	—	—

prosecutions.[17] In a study carried out in 1969 for the Law Commission,[18] it was found that proceedings under factories legislation were generally only instituted where the Inspectorate considered that the firm was at fault and that different

[17] Written Answers, Parl. Deb., H.C., March 2, 1978, cols. 295–6.
[18] Law Commission Working Paper No. 30 (*Strict Liability and the Enforcement of the Factories Act 1961*).

standards might be used as between enforcing remedial measures to avoid future accidents and for serious breaches of the law and accident cases where fault should be the criterion for prosecution. The gravamen of the study is that strict standards in the legislation are likely to be enforced on a "fault" basis.

Punishments

Trials on indictment in Crown Courts for the more serious offences have resulted in substantial fines. On November 27, 1979, Swan Hunter Shipbuilders Ltd. and Telemeter Installations Ltd. were fined £3,000 and £15,000 respectively, the latter being the highest recorded fine to date under the 1974 Act.[19] A fine of £10,000 was imposed at Gravesend Crown Court on Edmund Nuttall Ltd., civil engineers of London, who pleaded guilty to four charges following a lift accident in January 1978 at the Littlebrook 'D' Power Station in which four men were killed and five others were injured, the company as in the Swan Hunter Case having to pay costs.[20] Two companies were fined £16,500 for offences under the Asbestos Regulations of 1969 arising out of the release of clouds of asbestos dust following the removal of an old water tank and its lagging by scrap metal merchants who did not appreciate the danger in what they were doing.[21] Fines for offences tried summarily have come in for some general criticism and nowhere more than in the fines imposed for breaches of the 1974 Act. One trade union (SOGAT) has attacked the paltry fines imposed by magistrates' courts, a matter on which it seems to have secured the sympathetic ear of the Lord Chancellor, Lord Hailsham.[22] Figures disclosed by the trade union show that 316 convictions were secured against 217 persons in the four years from 1976 to 1979 and that the average fine was as follows—

Year	Max. Fine	Conviction	Average Fine
1976	£400	52	£ 99.42
1977	£400/£1,000	103	£106.69
1978	£1,000	89	£142.41
1979	£1,000	72	£165.90

19 See *The Fire on H.M.S. Glasgow*, September 23, 1976, H.M.S.O. 1978: *R.* v. *Swan Hunter Shipbuilders Ltd.* [1982] 1 All E.R. 264.
20 See *The Hoist Accident at Littlebrook 'D' Power Station* January 9, 1978, H.M.S.O.
21 *Construction News*, March 3, 1977, p.4.
22 The Magistrate (1982) 38, p. 17.

Time-limit

In general, proceedings for summary convictions must b
commenced within six months of the offence being committed,[2
but this general provision was modified by section 34 of the 197
Act, which allowed for an extension of time in certain cases, *e.g*
where a special report is made to which s.14(2)(*a*) applies, or
coroner's inquest is held. The time-limit of six months will ru
from the date when evidence justifying a prosecution came t
the knowledge of the responsible enforcing authority, provide
that there is a signed certificate from the latter stating the dat
when the required evidence came to its knowledge (sectio
34(3)). The six month time-limit however does not apply to an
indictable offence within the meaning of the Criminal Law Ac
1977, *i.e.* to summary proceedings for an offence triable eithe
way (see above), so that it follows that the certificate required b
section 34(3) is not required for the offences specified in sectio
34(4), which are offences triable either way. The time-limit fo
applying to a magistrates' court to state a case is 21 days.[24]

Venue

Section 35 provides that an offence under any of the relevan
statutory provisions, committed in connection with any plant o
substance, may be treated as having been committed at th
place where the plant or substance is for the time being, wher
this is necessary for the purpose of bringing the offence withi
the field of responsibility of any enforcing authority or conferring jurisdiction on any Court to entertain proceedings for th
offence.[25]

Fault of other person

Section 36(1) deals with a situation in which A has committed
an offence under Part I of the Act, or under the existing
statutory provisions, but the commission of that offence is due to
the act or default of B. In such a situation B may be charged with
and convicted of the offence whether or not proceedings are
taken against A. The other person (b) might be the safety officer.
Thus the safety officer who fails in his duty and causes a breach
of the Act can be charged as well as or instead of the employer.
The section does not provide a defence to A who remains liable.

Although the title of this section is "Offences due to fault of

[23] Magistrates' Courts Act 1980, s.127(1).
[24] M.C.A. 1980, s.111(2).
[25] See section 15(7), p. 112 above.

another person," the offence need not necessarily be an offence involving *mens rea* or negligence. B will be similarly liable where A is the Crown, which would itself otherwise have been liable (s.36(2)).

Who can prosecute?

Proceedings for an offence under Part I or the existing statutory provisions can only be instituted in England and Wales by an inspector or by or with the consent of the Director of Public Prosecutions (s.38). Therefore, if a person other than an inspector or the D.P.P. wishes to bring a private prosecution, the consent of the D.P.P. is needed. Ten applications for consent to proceedings had been received by the D.P.P. by the end of 1979. These applications came from local councils (4), the National Coal Board (1), private citizens (3) and Chief Officers of Police (2) of which only one (that from the Chief Officers of Police) was granted although the other from the same source was taken over by the D.P.P.[26] With regard to an offence which is to be prosecuted in the magistrates' court, an inspector, if authorised in that behalf by the enforcing authority which appointed him, may undertake the prosecution, although not counsel or a solicitor (s.39). In practice, the prosecution of minor offences on behalf of HSE is almost invariably undertaken by inspectors, although local authorities as enforcing authorities normally use their solicitors' department for this purpose. The competence of an inspector of the Health and Safety Executive to prosecute was successfully challenged in the magistrates' court in *Campbell* v. *Wallsend Slipway & Engineering Co. Ltd.*[27] on the ground that the inspector had not proved his competence by showing his own appointment and that of the Executive which had appointed him; but this finding was reversed on appeal to the Divisional Court, which held that the inspector did not need to be specifically empowered to prosecute under section 38 of the 1974 Act and did not need to produce a written instrument appointing the Executive under paragraph 20(3) of Schedule 2 to the Act, since that provision was directory only and not mandatory. The case exemplifies the maxim *omnia praesumunter rite esse acta* (all things are presumed to have been done correctly).

Offences by bodies corporate

In many cases under the 1974 Act, the "primary offender" in the case of a corporation is that artificial person—the corporate

[26] Written Answers, Parl. Deb., H.C., December 6, 1979. col. 322.
[27] [1977] Crim. L.R. 351.

body itself. Although it is still true that a corporation does not have a body to be hanged or a soul to be damned, it is now accepted without cavil that a corporation can commit a crime for which it may be charged. Because a corporation body is a legal abstraction incapable of doing a physical act or forming a mental attitude, the law accepts the inescapable, namely, that the corporation may only act through human agency. This human agency is not the *alter ego* of the corporate body but the directive mind of that body; the acts of the directive organs are in truth the acts of the corporation.[28] The acts or omissions of the directive organs are the acts or omissions of the corporation in a "personal" and not a vicarious sense.

In what is now a common-form provision, section 37 provides that where an offence under any of the relevant statutory provisions has been committed by a body corporate "with the consent or connivance of, or to have been attributable to any neglect on the part of, any director, manager, secretary or other similar officer of the body corporate or a person who was purporting to act in such capacity, he as well as the body corporate shall be guilty of that offence and shall be liable to be proceeded against and punished accordingly." The argument that the "neglect" referred to in section 37(1) relates only to a duty imposed on the defendant by the legislation was rejected in the Scottish case of *Armour* v. *Skeen* 1977 S.L.T. 70, where it was held that the director of roads for a local authority had neglected to implement *council* instructions on safety. The director in that case was convicted of an offence under the section, in addition to the council, which had failed in its duty under section 2(1) of the Act. The director was held to be a person covered by the phrase "any director, manager, secretary or other similar officer of the body corporate," which is clearly right, since he possessed considerable powers of direction and control. No doubt the purpose of section 37 is to warn those who act on behalf of a corporate body that they may be dragged out from behind "the corporate veil" and punished as individuals. However, without section 37 and its counterparts in other statutes, it ought not to be thought that individuals are inaccessible behind the corporate veil. Anyone, superior director or humble servant in the corporation, may be convicted for aiding and abetting the commission of an offence by the corporate body subject to the requirement that they must have *mens rea*. Conversely, the corporation itself may aid and abet the commission of an offence by its agent or servant.[29]

[28] See *Tesco Supermarkets* v. *Nattrass* [1972] A.C. 153.
[29] *R.* v. *Miller (Robert)(Contractors) Ltd.* [1970] 2 Q.B. 54.

There are arguments both ways as regards corporate criminal liability and the criminal liability of officers and servants of corporate bodies. The imposition of a fine on a corporation body, particularly one in the public sector, may result in punishment of the innocent to the extent that the cost of the fine is passed on to others, *e.g.* consumers. *Per contra*, corporations, particularly the larger corporations, are jealous of their corporate image and are anxious to avoid the public obloquy which a criminal conviction may bring upon them. The choice of defendant—corporation or individual—may on occasions pose an unenviable problem for a prosecutor. One of the themes in our health and safety law is that of individuated responsibility not only in connection with section 37 but also in connection with section 7 (duty of employees at work) and section 8 (interfering or misusing things provided pursuant to health, safety and welfare provisions).[30] Following upon the Public Inquiry into the causes and circumstances attending the explosion at Houghton Main Colliery on June 12, 1975, three officials at the colliery were found guilty of offences under the Mines and Quarries Act 1954 and Health and Safety at Work etc. Act 1974. They were fined a total of £1,750 and each was ordered to pay £200 towards the costs of the prosecution. On appeal to Sheffield Crown Court, however, the colliery manager was found not guilty of the two charges brought against him, although the appeal by the other officials (an under-manager and a deputy at the colliery) were dismissed, Judge Stephen saying that the prosecutions brought by the HSE were "only the tip of the iceberg."[31] A glance at the statistics dispels any notion that the Act has not been made to bear upon individuals as opposed to the corporation employing them. Thus some 60 informations were laid against officials and employees, as opposed to 702 against employers, for infringements of the 1974 Act in 1975, 1976 and 1077 (to June). Most of the former were laid under section 7 followed by informations laid under section 37.[32] More recent figures show that directors and managers were at the receiving end of only a handful of informations, considerably less than informations laid against employees which increased from two informations in 1975 to 38 informations in 1980. There is little doubt that the prosecution of individuals, particularly those who have assumed health and safety duties (possibly without extra payment), is a matter which has caused some disquiet amongst workers and their trade unions; a disquiet which HSE has sought to allay by

[30] These sections are discussed at pp. 101 *et seq*.
[31] Yorkshire Post, August 31, 1977.
[32] Written Answers, Parl. Deb., H.C., October 26, 1977, col. 875: see *ibid*. July 19, 1978, cols. 289–290 for details concerning different industries.

reassuring workers that only wilful and flagrant breaches will bring down the vengeance of the criminal law.[33]

An agreement under which an employer by means of a contract of employment or policy of insurance seeks to indemnify his employee against criminal liability under health and safety legislation is against public policy, at least for offences requiring guilty intent.[34] The safety representative is given immunity with regard to his functions as safety representative but this immunity does not extend to him *qua* employee.

CRIMINAL PROSECUTION AND CIVIL REDRESS

The Robens Committee was extremely critical of the manner in which statutes framed in terms of criminal law came to be used to found civil actions for damages arising out of personal injury and, indeed, came to be interpreted more commonly in the latter context than in the former. The civil action, when it lay at all, emphasised compensation at the price of prevention, introduced some refinements bordering on the sophistical, (as we have seen in Chapter 1), impeded the investigation of accidents and was costly, using up resources which might have been better employed in accident prevention.[35]

Section 47(1) segregates the criminal prosecution from the civil suit so far as Part I of the Act (which includes the "general duties") is concerned, by providing that nothing in Part I shall be construed as conferring a right of action in any civil proceedings in respect of any failure to comply with any duty imposed by sections 2 to 7 or any contravention of section 8, but this shall not affect the actionability of any breach of duty imposed by any of the existing statutory provisions or the operation of section 12 of the Nuclear Installations Act 1965 (which deals with compensation under that Act). Pending their replacement by regulations, the existing statutory provisions are actionable to the same extent, if any, as before. The Law Commission of England and Wales and, later, Lord Scarman, have called for explicit reference in protective statutes to the availability or non-availability of civil redress arising out of penal statutes and in recent statutes the matter has been taken out of the limbo of statutory construction with all of its uncertainties. The "general duties" provide no solace to those seeking treasures in what has been termed "the forensic lottery."

[33] Written Answers, Parl. Deb. H.C., February 19, 1982, cols. 251–253.

[34] In *Gregory* v. *Ford* [1951] 1 All E.R. 121, a lorry driver was allowed to recover the amount of a fine which he had to pay because his employer's vehicle did not carry compulsory traffic insurance.

[35] Report, paras. 434–438.

Section 47(2) is surprisingly explicit the other way in that, unless otherwise provided, it renders breach of a duty imposed by health and safety regulations civilly actionable, subject, of course, to damage, which is the gist of the action. It remains to be seen to what extent the regulations will follow the example of section 47(1) in excluding civil redress; certainly, the trade unions, whilst committed to accident prevention, are generally loth to see workers left with only their common law claims, minus any claim in respect of the tort of breach of statutory duty. Where the regulations provide for a specified defence in criminal proceedings,[36] that defence shall not be available in any civil proceedings, whether brought by virtue of section 47(2) or not, although as regards any duty imposed as mentioned in section 47(2), above, regulations may provide for any specified defence to be available in any action for breach of that duty.

If civil action is excluded by reason of section 47(1), or by the terms of regulations, such exclusion does not affect other rights of action, such as the action at common law for damages based upon negligence, or the defences available to such an action. Unless the regulations so permit, there can be no "contracting out" of any civil liability based upon regulations.

In *Groves* v. *Wimborne (Lord)*,[37] the Court of Appeal allowed a civil action for breach of section 3(4) of the Factory and Workshop Act 1878, despite a provision in that Act which permitted the Secretary of State to award part or whole of any fine imposed for breach of that section to the injured employee. The Robens Committee, as we have just seen, disliked the homologation of criminal and civil remedies in protective statutes. It is interesting to note that since the Powers of Criminal Courts Act 1873, sections 35–38, courts have a general power to make a compensation order requiring a person convicted of an offence to pay compensation for any personal injury, loss or damage resulting from that offence. The amount which a magistrates' court may award by way of compensation cannot exceed £1000,[38] but this differs from the statute construed in the Groves Case, in that the compensation order, whilst it cannot arise apart from a conviction, may be separate from any punishment awarded in respect of the offence, such as imprisonment or a fine. The different purposes of a fine and a compensation order might well be expressed in the words of Rigby L.J. in the Groves Case (who, was of course, dealing with an 1878 statute)—

[36] See s.15(6)(*b*).
[37] [1898] 2 Q.B. 402.
[38] Increased from £400 by Criminal Law Act 1977, s.60(1).

> "a very slight injury may be occasioned to a workman by a very gross and wilful neglect of the duty imposed by section 5. In such a case it would not be right for the magistrates to inflict a slight penalty, such as £5, because the injury caused was trivial. If they considered that there had been a deliberate evasion or neglect of the provisions of the Act, it might be right for them to inflict the full penalty of £100. But, on the other hand, there may be a case in which the injury to a workman is grievous, but the offence is comparatively venial. . . . It seems to me that there may be cases in which they would be entitled to say that the offence was slight though unfortunately it caused serious injury to the workman, and therefore only a small fine ought to be imposed."[39]

In making compensation orders, Courts are required to have regard to the means of the convicted person.

BURDEN OF PROOF

The burden of proving the commission of an offence under any of the relevant statutory provision rests upon the prosecution. The burden is higher than the civil burden which involves only proof on a balance of probabilities. The prosecution must persuade the court (or jury where applicable) of the defendant's guilt to the extent that it feels sure of his guilt to use the old time-honoured phrase. In a sense, both the criminal and civil standards of proof involve probabilities, except that the criminal standard involves what Denning J. referred to as "a high degree of probability," although not "beyond the shadow of a doubt" which would be too high an onus to place upon any prosecution.[40]

Some of the statutes prior to the 1974 Act, such as the Factories Act 1961 and its predecessors, were, in the words of Lord Upjohn "notoriously badly drafted,"[41] when it came to qualifying the burden of proof by some exception, proviso or the like; the qualifying phrase might be taken to be a matter to be proved by the accused, or, in some instances, it might be treated as part of the primary duty—"woven into the verb" of that duty, as one judge put it.[42]

Several of the statutory provisions require the accused to do something so far as is practicable, or so far as is reasonably practicable (a duty now given general currency by Part I of the

[39] S.C. pp. 414–415.
[40] *Miller* v. *Minister of Pensions* [1947] 2 All E.R. 372, 373–4.
[41] *Nimmo* v. *Alexander Cowan and Sons Ltd.* [1968] A.C. 107, 124.
[42] *Per* Lord Migdale in the Nimmo Case before the Court of Session.

1974 Act), or to use the best practicable means to achieve some state of affairs. Section 40 makes it clear that it is for the accused to prove that these requirements have been met. It is not correct to say that the whole burden is upon the accused, or to say that the burden "shifts" to him; the prosecution, notwithstanding section 40, must discharge its own burden, *e.g.* that the accused has not provided a "safe means of access" to a place of work under section 29 of the Factories Act 1961. It would then be incumbent upon the accused to demonstrate that he had done everything which was "reasonably practicable." The phrase in section 40, "it shall be for the accused to prove," places upon him the legal or persuasive burden of proof on a balance of probabilities; not the higher burden placed on a prosecution (beyond reasonable doubt) discussed above.

Relevance of breach of general duties in subsequent civil proceedings.

It has already been seen above that the general duties are supported by the criminal sanction alone and do not provide any basis for civil action. Section 11(1) of the Civil Evidence Act 1968 provides that:

> "In any civil proceedings the fact that a person has been convicted of an offence by or before any court in the United Kingdom or by a Court Martial there or elsewhere shall be admissible in evidence for the purpose of proving, where to do so is relevant to any issue in those proceedings, that he committed that offence but no conviction other than a subsisting[43] one shall be admissible in evidence by virtue of this section."

This section reverses the much-criticised common law rule in *Hollington* v. *Hewthorn*[44] which precluded the use of a criminal conviction in civil proceedings and implements the recommendation of the Fifteenth Report of the Law Reform Committee.[45]

Section 11(2) of the Civil Evidence Act 1968 further provides:—

> "In any civil proceedings in which by virtue of this section a person is proved to have been convicted of an offence by or before any court in the United Kingdom or by a Court Martial there or elsewhere—

[43] If not subject to or quashed on appeal: Re *Raphael (deceased)* [1973] 3 All E.R. 19.
[44] [1943] K.B. 587.
[45] Cmnd. 3391, Chairman Rt. Hon. Lord Pearson, C.B.E.

(a) he shall be taken to have committed that offence unless the contrary is proved; and

(b) without prejudice to the reception of any other admissible evidence for the purpose of identifying the facts on which the conviction was based, the contents of any document which is admissible as evidence of the conviction, and the contents of the information, complaint, indictment or charge sheet on which the person in question was convicted shall be admissible in evidence for that purpose."

These provisions may to some extent have thwarted the intention of Parliament in segregating criminal and civil liabilities in relation to the general duties. The intentions of the Government, as expressed in Committee Stage of the Bill,[46] were to ensure that nothing would conflict with established practice; the existing relevant statutory provisions were to be unaffected and no new liabilities for personal injuries in civil proceedings were to be created pending the outcome of the Pearson Commission.[47]

It may be that employers when faced with a prosecution under the general duties will be less ready to plead guilty than would have been the case had section 11 of the Civil Evidence Act 1968 not been enacted, inasmuch as a conviction may be used in subsequent civil proceedings, a consideration to be borne in mind in respect of the eight million or so "new entrants," who were outside the ambit of health and safety legislation prior to January 1, 1975, but who are now covered by the general duties in Part I of the 1974 Act.

Remedying the cause of the offence and forfeiture: section 42

Dissatisfaction with the punitive role of the criminal law has caused attention to be turned to other social purposes which Courts might take into account when determining what orders they should make. Restitution, compensation and community service are examples of this new and enlightened trend, of which section 42 is another example in the field of health and safety at work. If the Court considers that a person convicted of an offence under any of the relevant statutory provisions has it in his power to remedy any matters relating to that offence, it may order him to take steps to remedy those matters, either in addition to or instead of an order for punishment. Where time

[46] Official Report, Standing Committee A, Health and Safety at Work Bill, May 14, 1974, col. 244 (Mr. Harold Walker).

[47] This Commission reported in January 1978, Cmnd. 7054.

for remedying matters has been fixed by the Court order, the Court may on application before such time has elapsed, extend or further extend the time for compliance with the order. During any time given for purposes of compliance, including extensions of time, the person convicted is not to be liable under any of the relevant statutory provisions in respect of those matters.

In those cases in which a person is convicted of an offence relating to the acquisition, attempted acquisition, possession or use of an explosive article or substance in contravention of any of the relevant statutory provisions, the Court may order that article or substance to be forfeited, destroyed or otherwise dealt with, unless, on application to be heard, by a person claiming to be the owner of such article or substance an opportunity has not been given to the owner or alleged owner to show cause why the order should not be made.[48]

Failure to comply with a court order under section 42 is an offence by virtue of section 33(1)(*o*), triable summarily or on indictment (in which latter event imprisonment for a term not exceeding two years is a possibility under section 33(3)(*b*)(i)).

[48] The Explosives Act 1875, save for certain sections thereof, is listed in Schedule 1 to the 1974 Act as an existing enactment which is one of the "relevant statutory provisions." The Explosives Inspectorate brought only one prosecution in 1980 and that was against an employee for breach of section 7 of the 1974 Act. The Inspectorate is empowered in appropriate circumstances to grant exemptions in relation to the operation of the 1875 Act: HSE Report for Manufacturing & Service Industries 1980, H.M.S.O. 1982.

CHAPTER 7

OBTAINING AND DISCLOSURE OF INFORMATION

The philosophy of disclosure

It is axiomatic that any effective policy for health and safet at work depends upon obtaining and communicating sufficien information. The "philosophy of disclosure" is fundamental to policy dedicated to the changing of attitudes. Research at hom and abroad, including the research carried out by the Researc and Laboratories Services Division of HSE, produces a constan flow of new data which must be evaluated before it can be passe on to employers and workers. Research may disclose a risk in particular substance or process (not necessarily with health an safety at work in mind) but the risk may be imprecise an difficult to translate into specific guidance. In the United State of America, the Occupational Safety and Health Administratio (OSHA) (the American counterpart to our HSC/HSE) has bee castigated by the Supreme Court for promulgating an asbesto standard which is pitched at a conservatively safe level in th absence of absolutely firm knowledge of what the safe level i Because perfect knowledge is hard to come by in this imperfec world, our authorities have to strike a balance between givin inadequate warning of a risk and causing unnecessary an unjustified alarm. HSC is obliged by section 11(2)(*b*) to carry ou and publicise research and to encourage others to carry ou research and to communicate information. HSC/HSE produce regular flow of information in sector reports, committee report research papers, guidance notes, special guidance notes, discus sion documents, advisory leaflets, booklets and in regulation themselves. Several of the latter are based on the need fc information, *e.g.* the Notification of Accidents and Dangerou Occurrences Regulations 1980 (discussed at p.217) or the Safet Signs Regulations 1980. The information derived under th former Regulations is now fed into a national computerised dat storage and retrieval system known as SHIELD which wi permit HSE to process information and other information suc as complaints, visits by inspectors, legal enforcement notic and prosecutions. Regulations are proposed for potentiall hazardous installations under which the occupiers of thos installations will be obliged to give information of the proposed activities.

It is perhaps disingenuous to assume that we are all committed to the free dissemination of knowledge. Max Weber prophesied that timidity and secrecy would be *par excellence* the hallmarks of the new bureaucracies. The effects of lead in petrol on children, the conveyance of nuclear waste and the dumping of toxic wastes arouse fears on the part of people who are better informed and more demanding than their forbears. Similarly, manufacturers and suppliers of products are not in business to communicate information. The larger employer may be fully aware of the benefits of investment in health and safety measures as revealed in better industrial relations, reduced absenteeism and increased profits but, regrettably, a few of the smaller employers distinguish between "ought" and "is" fearing that the imparting of information to their workers may be tantamount to giving a hostage to good fortune.

In what follows, the obtaining and disclosing of information will be treated in relation to (a) the enforcing authorities, and (b) employers and their employees.

THE ENFORCING AUTHORITIES

Investigative powers of inspectors

Amongst the wide powers given to inspectors by section 20 to enable them to carry into effect the relevant statutory provisions within the field of responsibility of his enforcing authority are those—

- to enter premises in order to make examinations and investigations,
- to take measurements and photographs,
- to take samples of articles or substances (or to take possession of the same),
- to require any person whom he has reasonable cause to believe to be able to give any information relevant to any examination or investigation (above), and
- to inspect and take copies of books or documents.

An important qualification of these wide powers is imposed by section 20(8) which states that nothing in section 20 shall be taken to compel the production by any person of a document of which he would on grounds of legal professional privilege be entitled to withhold production on an order for discovery in an action in the High Court, or, as the case may be, on an order for the production of documents in an action in the Court of Session of Scotland.

In other words, a person such as an employer may refuse to disclose a document to an inspector if he could refuse to disclose that document on an order for discovery in a High Court action provided that the refusal is covered by "legal professional privilege." Under Order 24, rule 1, Rules of the Supreme Court, it is possible to obtain discovery by the parties to an action begun by writ of documents which are or have been in question in the action. This right of discovery does not extend to documents for which privilege, of which legal professional privilege is one form, is established under Order 24, rule 5. Legal professional privilege attaches to documents involving private communications between a solicitor and his client (or between a company and its legal adviser) quite apart from the question of whether or not litigation is contemplated or pending. It also attaches to documents which were made or which came into existence at a time when litigation was contemplated or pending. It often happens that a document comes into existence for two purposes, such as to explain the causes of an accident as found by an inquiry and to ascertain legal liability in connection with any litigation arising out of that accident. It was formerly considered that legal professional privilege could extend to such a "dual-purpose" document, a view expressed by Viscount Dilhorne (dissenting) in a 1974 case—

> "Where an event occurs which is likely to lead to litigation, *e.g.* an accident on a railway, it has long been established that reports made in anticipation of litigation and for the use of the defendant's solicitors are protected, and that the reports need not be made solely or primarily for the use of the solicitors."[1]

Following a review of the law in *Waugh* v. *British Railways Board*,[2] the House of Lords overruled the previous authorities[3] and held that unless the sole or dominant purpose for which the document was prepared related to litigation contemplated or pending, privilege should not attach to the document. The "dual-purpose" document must be disclosed if requested, and whilst it is true that the safety of persons might well depend on the candour and completeness of accident reports this of itself is not sufficient to establish legal professional privilege.

Obtaining information by HSC, HSE, enforcing authorities

In addition to the inspector's information-seeking powers under section 20, discussed above, section 27 confers upon HSC

[1] *A. Crompton Ltd.* v. *Customs and Excise Cmrs. (No. 2)* [1974] A.C. 405, at p. 421.
[2] [1979] 2 All E.R. 1169.
[3] *Birmingham and Midland Motor Omnibus Co. Ltd.* v. *London and North Western Railway Co.* [1913] 3 K.B. 850; *Arkin* v. *London and North Eastern Railway Co.* [1929] All E.R. Rep. 65; *Ogden* v. *London Electric Railway Company* [1933] All E.R. Rep. 896.

the role of potential Torquemada with regard to any information which it needs for the discharge of its functions or which an enforcing authority needs for the discharge of its functions. With the consent of the Secretary of State (for Employment) (which may be given generally for cases of any stated description), it may serve on any person a notice requiring that person to furnish to it (or the enforcing authority) such information about such matters as may be specified in the notice, and to do in such form and manner and within such time as may be so specified. A great deal of information is available to governmental bodies other than HSC and section 27(2) allows this information to be "tapped" by HSC or HSE. The subsection provides that nothing in section 9 of the Statistics of Trade Act 1947 (which restricts the disclosure of information obtained under that Act) shall prevent or penalise—

(a) the disclosure by a Minister of the Crown to HSC or HSE of information obtained under that Act about any undertaking being information consisting of—
   - the names and addresses of persons carrying on the undertaking,
   - the nature of the undertaking's activities,
   - the numbers of persons of different descriptions who work in the undertaking,
   - the addresses or places where activities of the undertaking are or were carried on,
   - the nature of the activities carried on there,
   - the number of persons of different descriptions who work or worked in the undertaking there;

(b) the disclosure by the Manpower Services Commission[4] to the HSC or HSE of information so obtained which is of a kind specified in a notice in writing given to the disclosing body and the recipient of the information by the Secretary of State under this paragraph.

The requirement in section 27(4) that information disclosed to a person in pursuance of (a) or (b), above, shall not be used for a purpose other than that of HSC or HSE reflects a proper anxiety concerning possible misuse of information in an age of data banks and computerised information retrieval. Sometimes the information held by one Government Department is considered to be so confidential that it will not be released to another Government Department. An expensive Prices and Incomes Board was set up in the 1960s to acquire information much of which was already known to the Board of Inland Revenue. Section 27(3) includes in references to a Minister of the Crown,

[4] The Employment Service Agency and the Training Services Agency, both mentioned in section 27(2)(*b*), were abolished by section 9, Employment and Training Act 1981.

HSC, HSE and MSC, individuals in those bodies such as officers. Information obtained under section 27 must only, as has been noted, be used for a purpose of HSC or HSE. Section 28 places restrictions on the disclosure of information.

Restrictions on disclosure

Subject to some important exceptions, section 28 forbids the disclosure of "relevant information" without the consent of the person by whom it was furnished. "Relevant information" means information obtained by HSC under section 27(1) (above) or furnished to any person in pursuance of a requirement imposed by any of the relevant statutory provisions. This general restriction on disclosure does not, however, apply to the following exceptions which are set out in section 28(3), namely—

(a) disclosure of information to the HSC, HSE, Government Department or any enforcing authority;
(b) without prejudice to (a), disclosure by the recipient[5] of information to any person for the purpose of any function conferred on the recipient by or under any of the relevant statutory provisions;
(c) without prejudice to (a), disclosure by the recipient to an officer of certain authorities (local authority, water authority or water development board, river purification board) which authorise him to receive it, or a constable authorised to receive it by a chief officer of police;
(d) disclosure by the recipient of information in a form calculated to prevent it from being identified as relating to a particular person or case;
(e) disclosure for the purposes of any legal proceedings or any investigation or inquiry held by under section 14(2) (this deals with the power of HSC to direct investigations and inquiries) or for the purposes of a report of any such proceedings or inquiry or of a special report made by virtue of section 14(2).

References to HSC, HSE, a Government Department or enforcing authority include references to their officers, or, where applicable, their inspectors and, in the case of a reference to the HSC, bodies and their officers with whom HSC or HSE has an agency agreement. The five categories in (a) to (e) above do not receive *carte blanche* with the relevant information and must not use it for a purpose other than—

[5] Defined in relation to any relevant information as the person by whom that information was so obtained or to whom that information was so furnished, as the case may be (s.27(1)).

(a) in a case falling within paragraph (*a*), a purpose of HSC, HSE, Government Department, or the enforcing authority in connection with the relevant statutory provisions, as the case may be;
(b) in the case of information given to an officer of a local authority, water authority, river purification board or water development board, the purposes of the authority or board in connection with the relevant statutory provisions or any enactment whatsoever relating to public health, public safety or the protection of the environment;
(c) in the case of information given to a constable, the purposes of the police in connection with the relevant statutory provisions or any enactment whatsoever relating to the public health, public safety or the safety of the State.

With regard to information obtained as a result of an inquiry under section 14(4)(*a*), or the exercise of inspectorial powers under section 20 (with particular reference to trade secrets), the person so acquiring the information shall not disclose it, except for the purposes of his functions, or for the purposes of any legal proceedings or investigation or inquiry under section 14(2), or for the purposes of a report of such proceedings or inquiry, or special report under section 14(2), or with the relevant consent. In this connection "relevant consent" means, in the case of information acquired under section 20, the consent of the person who furnished it, and, in any other case, the consent of a person having responsibilities in relation to the premises where the information was obtained.

Section 28(8) dispenses with any need of "relevant consent" in those circumstances in which an inspector may reveal certain information where it is necessary to do so for the purpose of assisting in keeping persons (or the representatives of persons) employed at any premises adequately informed about matters affecting their health, safety and welfare. The information which may be given is either—

(a) factual information obtained under section 14(4)(*a*) or section 20 which relates to the premises, anything done (or being done) therein; and
(b) information with respect to action taken or to be taken in connection with those premises in the performance of his functions.

In either event, the inspector must also give the like information to the employer of the employed persons or their representatives. In practice, of course, this disclosure by inspectors will be revealed to safety representatives, who will, in practice, be informed of improvement and prohibition notices served on their employer.

An important extension of the power to disclose information is to be found in subsection (9) of section 28. A person who has obtained information as the result of an inquiry, or as the result of the exercise of an inspector's power under section 20, may furnish to a person who appears to him to be likely to be a party to any civil proceedings arising out of any accident, occurrence situation or other matter, a written statement of relevant facts observed by him in the course of exercising any of the powers incidental to an inquiry or under section 20.[6]

Authorities and private litigation

The findings and report of an inspector who has visited the site of an accident are likely to be of direct relevance to a claim for personal injury arising out of that accident. In its evidence to the Robens Committee, the Department of Employment related its experience, which was that—

> "the possibility of claims for damages sometimes hampers enforcing authorities in investigating the circumstances and causes of accidents or dangerous occurrences. This happens because those concerned, whether managers or workers, are more inhibited than they might otherwise be in disclosing what happened, who was doing what, what the works instructions were, etc."[7]

Understandably, the enforcing authorities are more concerned with the prevention of accidents than in providing evidence for the use of private litigants. The authorities wish to obtain compliance with the law, using primarily persuasion, but falling back where necessary on administrative notices and, as an ultimate sanction, prosecution. By becoming caught up in the toils of civil litigation they are distracted from their preventive and enforcement roles. Section 28(9), discussed above, gives a *discretion* to furnish a written statement; there is no *duty* to give such a statement.

We have seen that it is possible for litigants to obtain documents prepared for the use of the inspectorates, or for an employer by his legal adviser, provided that it is not the *dominant* purpose of those preparing the document or report that it should be submitted to a legal adviser for advice and use in litigation. Material which comes into existence for more tha

[6] The fuller the information in the written statement, the more likely that a potential claimant wi not press for fuller disclosure under rules of court (below).
[7] Vol. 2, Report of the Committee, p. 284, and see the Report itself at para. 436.

one purpose, *e.g.* to ascertain the cause of an accident and to provide information for a legal adviser, must be disclosed if requested.[8]

Section 31 of the Administration of Justice Act 1970 deals with parties who appear to the High Court to be likely parties to a personal injuries (or fatality) claim and empowers the High Court to order disclosure of documents. This section deals with the parties to a possible personal injuries (or fatality) action. Section 32 goes further in that it enables the High Court to order disclosure against "a person who is not a party to the proceedings and who appears to the court to be likely to have or to have had in his possession, custody or power any documents which are relevant to an issue arising out of that (*i.e. personal injuries or fatality*) claim" (*italics supplied*). Such an order could issue against HSE in respect of documents relevant to such a claim. This extended power to obtain discovery from third parties (with us the enforcing authority) does not apply before the commencement of proceedings, unlike the power conferred by section 31. The practice for using section 32(1) is set out in the Rules of the Supreme Court, Order 24, rule 7A, and would, for example, be particularly useful in a claim involving toxic hazards where the HSE has visited the workplace over many years in connection with a hazard of this description.[9]

HSC, HSE, their officers and servants perform their functions "on behalf of the Crown,"[10] and at one time could refuse to disclose information on the ground that the disclosure would be "injurious to the public interest." It is now recognised that the "public interest" works in two directions. An investigation into an accident to a nuclear submarine may well raise a public interest in non-disclosure on the basis of national security, whereas the findings of an investigation into an accident in a factory is more likely to point to the public interest in disclosing those findings. The scope of refusal on grounds of public interest has been drastically curtailed by the House of Lords decision in *Conway* v. *Rimmer*,[11] where Lord Reid discounted the argument that the public safety might be affected if the candour and completeness of accident reports by public bodies were to be inhibited by reason of the possibility that the reports might have to be produced in the interests of the due administration of justice. Lord Pearce in the same case pointed to the "extreme malaise" which might result from a total acceptance of the theory that all documents, such as police reports of accidents,

[8] See above, p. 180.
[9] See further Legal Action Group *Bulletin* 1976, p. 59.
[10] Section 10(7) of the 1974 Act.
[11] [1968] A.C. 910.

should be protected in the public interest. The tide in favour of discovery has reached the industrial tribunals in discrimination cases where the test used is whether or not discovery is needful in order to dispose fairly and justly with the case.[12]

EMPLOYERS AND WORKERS

It is now generally recognised that it is good industrial relations practice for management to disclose information to trade union representatives so that effective collective bargaining can take place. The Employment Protection Act 1975, sections 17 to 21,[13] repeating similar provisions in the Industrial Relations Act 1971 (never activated), requires an employer (subject to some important exceptions) to disclose on request to representatives of an independent trade union information relating to his undertaking which is in his possesion and which, if withheld, would impede collective bargaining to a material extent and would not, at the same time, conform to good industrial relations practice. The duty only appertains in relation to a recognised trade union and has been amplified by a Code of Practice which came into operation in 1977.[14] This statutory duty of disclosure is intended to coerce the unduly uncommunicative employer who does not subscribe to the "philosophy of disclosure," a philosophy now generally accepted by Government, management and labour. If health and safety is, as some would maintain, a matter for collective bargaining, it could have been brought within the ambit of this general statutory disclosure. However, as we saw in Chapter 1,[15] health and safety at work is treated, not as a matter for conflict to be resolved by collective bargaining, but as involving a community of interests for which joint consultation is the appropriate mechanism. The machinery of safety representatives and safety committees is discussed elsewhere in this work,[16] but it will be convenient to deal here with the manner in which the philosophy of disclosure is intended to operate in relation to that machinery (which is essentially consultative).

Before doing so, mention might be made of a provision added to the 1974 Act as it went through Parliament, namely, section 79, which amends the Companies Act 1967 by empowering the

[12] *Science Research Council* v. *Nassé* [1979] 3 All E.R. 673.
[13] The procedure to obtain disclosure is cumbersome. Disclosure of information has not been as controversial as it might have been. The CBI guide issued some years ago was similar in many respects to the CIR guidelines under The Industrial Relations Act 1971.
[14] There is a duty to consult representatives of a recognised trade union in connection with impending redundancies under s.99, Employment Protection Act 1975.
[15] See pp. 30 *et seq.*, *supra.*
[16] See Chapter 8, pp. 190 *et seq.*, *infra.*

Secretary of State to make regulations concerning the addition of health and safety and welfare arrangements for employees of the company and others to the matters to be dealt with in directors' reports. No such regulations have been made at the time of writing. This new provision is based upon the view that employees have as much (some would say more) right to information concerning their company as shareholders, whose involvement in the company, apart from ownership of shares and receipt of dividends, may be small or non-existent. The system of safety representatives places the safety representatives between management and the workforce.

Safety representatives: Inspection of documents and provision of information

Apart from their general functions and powers, safety representatives have positive powers to acquire information in addition to the more usual right to receive information. Regulation 7 of the Safety Representatives and Safety Committees Regulations 1977 provides in forthright terms that—

> (1) . . . Safety representatives shall for the performance of their functions under section 2(4) of the 1974 Act and under these Regulations, if they have given the employer reasonable notice, be entitled to inspect and take copies of any document relevant to the workplace or to the employees the safety representatives represent which the employer is required to keep by virtue of any relevant statutory provision within the meaning of section 53(1) of the 1974 Act except a document consisting of or relating to any health record of any identifiable individual.

So far as disclosure is concerned, the same Regulation requires that—

> (2) An employer shall make available to safety representatives the information, within the employer's knowledge, necessary to enable them to fulfil their functions.

The employer is not bound, however, to make known the following classes of information—

(a) any information the disclosure of which would be against the interests of national security; or
(b) any information which he could not disclose without contravening a prohibition imposed by or under an enactment; or
(c) any information relating specifically to an individual, unless he has consented to its being disclosed; or

(d) any information the disclosure of which would, for reasons other than its effect on health, safety or welfare at work, cause substantial injury to the employer's undertaking or, where the information was supplied to him by some other person, to the undertaking of that other person; or

(e) any information obtained by the employer for the purpose of bringing, prosecuting or defending any legal proceedings.

(3) Paragraph (2) above does not require any employer to produce or allow inspection of any document or part of a document which is not related to health, safety or welfare. I the withholding of such a document would impede collective bargaining it may be disclosable under section 17 of the Employment Protection Act 1975 (see above).

Paragraph 6 of the Approved Code of Practice on Safety Representatives and Safety Committees particularises the sorts of information which should be made available to safety representatives by the employer and which should include—

(a) information about the plans and performance of thei undertaking and any changes proposed insofar as the affect the health and safety at work of their employees;

(b) information of a technical nature about hazards to healtl and safety and precautions deemed necessary to eliminat or minimise them, in respect of machinery, plant, equip ment, processes, systems of work and substances in use a work including any relevant information provided b consultants or designers or by the manufacturer, importe or supplier of any article or substance used, or proposed t be used, at work by their employees;

(c) information which the employer keeps relating to th occurrence of any accident, dangerous occurrence c notifiable industrial disease any statistical records rela ing to such accidents, dangerous occurrences or cases notifiable industrial disease;

(d) any other information specifically related to matte affecting the health and safety at work of his employee including the result of any measurements taken by th employer or persons acting on his behalf in the course checking the effectiveness of his health and safet arrangements;

(e) any information on articles or substances which a employer issues to homeworkers.

We have already seen that health and safety inspectors ar

subject to a positive duty under section 23(8) of the 1974 Act to communicate certain matters affecting the health, safety and welfare of employees to the employees or their representatives.[17] In practice, the disclosure of information will be to the safety representative, who thus becomes entitled to information, both from the employer and the inspector. That, at least, is the theory. If the safety representaive finds himself confronted with an obstructive employer or a Trappist-like inspector there is no directly available legal remedy which he can invoke. The refusal or failure of the employer to discharge his duties under the Regulation may give rise to a collective dispute to be settled by using procedure (if any), by resort to ACAS for conciliation or arbitration or, possibly, industrial action. The safety representative is at liberty to complain to the inspector whom he may attempt to prevail upon to issue an Improvement Notice to enforce the Regulations, or to commence a prosecution; but if the inspector declines to be stirred into either form of action, there is little that the representative can do: he cannot, as a private citizen, obtain a *mandamus* against the inspector to carry out his duty under section 28(8), leaving on one side the discretion given to the inspector under that provision in determining whether or not it is "necessary" to give the information. Because, as we have seen, health and safety is not seen as a matter for collective bargaining, the tortuous procedure for obtaining disclosure as set out in the Employment Protection Act 1975 would appear to be inapplicable. It could be, however, that health and safety, whilst not a subject for collective bargaining, might be *relevant* to such bargaining, *e.g.* where a rate of "danger money" is being negotiated and information germane to that matter is in the possession of the employer. We have seen that if legal proceedings are taken against the employer by an employee, *e.g.* for personal injuries, discovery of documents is available to a party, not only against the defendant but also against a third party.[18]

[17] See above p. 183.
[18] See above p. 185.

CHAPTER 8

SAFETY REPRESENTATIVES AND SAFETY COMMITTEES

The Background

The Webbs have described how the worker in the nineteenth century submitted to squalid and even dangerous conditions of work because control of the working environment was seen as essentially a concomitant of the employer's ownership of the enterprise. The "wages-contract" provided only for wages and hours of work leaving the conditions in which the work was to be performed as a matter of managerial privilege. Indeed, to this day, the conditions of work almost invariably lie outside the scope of the individual contract of employment if only because it would be impossible to include them as contractual terms between employer and employee. It was left to the trade unions to protect their members by negotiating improved conditions of work quite apart from improved terms relating to pay and hours of work. We have already mentioned[1] the way in which the trade unions themselves , whilst often capable of improving conditions of work for their members, later sought more uniform solutions in the form of Acts of Parliament, of which factories legislation was the first example, although that legislation owed more to enlightened reformers than to pressure from what were then embryonic trade unions. The gap left by the individual contract of employment was filled by collective bargaining and by legislation. Section 29(1) of the Trade Union and Labour Relations Act 1974 defines a "trade dispute" as a dispute connected with certain matters, of which the first and most important, are "terms and conditions of employment or the physical conditions on which any workers are required to work."[2]

Although this contractualist approach to health and safety has never disappeared as an element in "oppositional" bargaining, it has been increasingly supplemented by statutory standards enforced by governmental agencies. The Robens Committee, as we have noted,[3] saw health and safety as a matter of

[1] See Chapter 1 above, at pp. 32 *et seq.*
[2] Also the main ingredient in a collective agreement as defined by s.30(1) of the same Act. The collective agreement and the trade dispute may relate to "facilities for officials of trade unions" which could include facilities for safety representatives appointed by trade unions.
[3] Discussed in Chapter 1, at pp. 33 *et seq.*

"common interest" between management and workers to be resolved at the workplace by proper consultation. The high hopes of the Seventies that there would be structual changes in the composition of enterprises in Great Britain portended by the long-awaited Fifth Draft Directive of the EEC, and the appointment of the Bullock Committee on Industrial Democracy,[4] were replaced by apathy when the Fifth Directive failed to appear and the Bullock Report was consigned to Whitehall archives. The latter report gave a peculiarly British twist to the continental forms of "co-determination" by insisting on the paramount role of trade unions in any new company structures. There has in recent decades been a fundamental difference between those who espouse *worker participation* and *union-sponsored worker participation*, a difference which was graphically illustrated in the somewhat acrid debates on the dropping of section 2(5) from the 1974 Act by an amendment in 1975.[5] Section 2(5) would have permitted the Secretary of State to make regulations whereby employees themselves could have elected their own safety representatives. This leaves us with section 2(4) which states—

> "Regulations made by the Secretary of State may provide for the appointment in prescribed cases by recognised trade unions (within the meaning of the regulations) of safety representatives from amongst the employees, and those repesentatives shall represent the employees in consultations with the employers under subsection (6) below and shall have such other functions as may be prescribed."

We are left, therefore, with union-sponsored worker participation through safety representatives, now amplified in Regulations with an accompanying approved Code of Practice and Guidance Notes.

The system of safety representatives now introduced into industry (including the public sector of industry) owes much to the experience of joint regulation of safety (not to mention checkweighing) in the mines. Because of the particular dangers in coal mines, such as gas or roof falls, miners obtained in General Rule 30 of the Coal Mines Regulation Act 1872 the right to appoint two of their number to inspect the pit once a month, a right improved by the Coal Mines Act 1911, section 16, and now regulated by section 123 of the Mines and Quarries Act 1954. Joint regulation of safety in the mines was not, however, mirrored in the rest of British industry for even by 1970 less

[4] Report of the Committee of Inquiry on Industrial Democracy (Chairman Lord Bullock) (1977): Cmnd. 6706.
[5] Employment Protection Act 1975, ss.116, 125(3) and Schedule 15, para. 2 and Schedule 18.

than half of the factories employing fifty persons or more hac safety committees. Prior to the 1974 Act, management-unior consultation on health and safety was generally poor.

Appointment of safety representatives

The Health and Safety Commission prepared proposals for th appointment of safety representatives and safety committees but these encountered some criticism[6] based mainly on th likely financial implications of the new system so that, althoug the regulations[7] were made on March 16, 1977, their imple mentation (and that of the accompanying approved code o practice and guidance notes) was delayed until October 1, 1978. TUC representatives saw the Prime Minister on two occasion concerning the delay in the implementation of the regulations.

Appointment of safety representatives. The regulations allow recognised trade union to appoint safety representatives fron amongst the employees in all cases where one or mor employees are employed by an employer by whom that union i recognised, except in the case of employees employed in a min as defined in section 180 of the Mines and Quarries Act 195 which is a coal mine. The regulations define a "recognised trad union" as—

> "an independent trade union as defined in section 30(1) c the Trade Union and Labour Relations Act 1974 which th employer concerned recognises for the purpose of negoti tions relating to or connected with one or more of th matters specified in section 29(1) of that Act in relation t persons employed by him."

In order to have the right to appoint safety representatives, th union must (1) be a trade union as defined in section 28(1) TULRA 1974 (which would exclude a committee of sho stewards acting as a temporary "ginger group" or a pure professional association such as the Law Society), and (2) b "independent." To be independent a trade union must (a) not b under the domination or control of an employer or a group employers or of one or more employers' associations (th "company-dominated union" of American labour law) and (not be liable to interference by an employer or any such group

[6] Particularly on the part of local authorities: see, for example, *Municipal Review* No. 558, Ju 1976, p. 75, expressing fears over resource problems.

[7] Safety Representatives and Safety Committees Regulations 1977 S.I. 1977 No. 500 (publish together with a Code of Practice and Guidance Notes).

[8] See Parl. Deb. vol. 923, November 19, 1976, col. 797 (Written Answers); vol. 924, February 1, 19 col. 153 (Written Answers).

[9] Financial Times, February 3, 1977, p. 16.

association (arising out of the provision of financial or material support or by any other means whatsoever) tending towards such control.[10] A trade union is "liable to interference" if it is vulnerable to the risk of interference and it is now necessary to show that it is likely to be subjected to such interference.[11] A listed trade union may apply to the Certification Officer under section 8 of the Employment Protection Act 1975 for a certificate that it is independent (which is conclusive on that question whilst it is in force). A trade union aggrieved by the refusal of a certificate of independence (*not* a trade union aggrieved by the grant of a certificate to another union) may appeal on law or fact to the Employment Appeal Tribunal by virtue of section 88(3) of the 1975 Act.

The trade union must also be "recognised" for the purpose of negotiations relating to or connected with the seven matters on which collective bargaining can, legally speaking, take place (see section 29(1) of TULRA 1974).[12] Recognition may be (a) express or (b) implied from conduct, provided that the conduct in question is sufficiently unequivocal to demonstrate a clear intention on the employer's part and not some equivocal conduct such as allowing a trade union official to put a notice on the employer's notice board or to recruit members in the workplace. The Guidance Notes, paragraph 5 state that "The Commission have decided that recognition for this purpose must be on the same basis as in the Trade Union and Labour Relations Act 1974 and the Employment Protection Act 1975"[13] and, also, that "any disputes between employers and trade unions about this matter should be dealt with under the provisions of the Employment Protection Act." The cases decided under this latter Act turn upon two parts of this latter Act, *viz.* recognition simpliciter (now stripped of the statutory recognition procedures repealed by the Employment Act 1980) and the duty to consult recognised trade unions in connection with impending redundancies. In *USDAW* v. *Sketchley Ltd.*,[14] an agreement between the employer and the trade union, termed a "recognition for representation agreement," *i.e.* to allow the union to take up limited representational duties in connection with grievances and to allow for the appointment of shop stewards and the collection of union dues did not raise the inference that

[10] *Ibid.* s.30(1).
[11] *Squibb* v. *United Kingdom Staff Association* [1979] I.C.R. 235.
[12] The repeal of the statutory recognition provisions in sections 11–16 of the Employment Protection Act 1975 by section 19 of the Employment Act 1980.
[13] The nationalisation statutes contain loosely-worded obligations to consult concerning recognition. See on s.11, Post Office Act 1969, *ASTMS (TCO Section)* v. *Post Office* [1980] I.R.L.R. 475.
[14] [1981] I.R.L.R. 291.

recognition for purposes of negotiation had been impliedly granted.

Once the trade union has appointed the safety representative in accordance with any rules or custom applicable to such appointments, it or a person acting on its behalf must notify the employer in writing of the appointee's name and the group or groups of workers whom he (or she) will represent. It is important that a safety representative should be properly appointed to forestall possible attacks on his credentials by fellow union members or others which might end up in the tribunals or courts. The employer might, for example, challenge an appointment by refusing to grant time-off to the appointee to enable him to carry out the functions which we describe below. Under the regulations, the trade union may revoke the appointment of a safety representative but must notify the employer in writing of the termination of the appointment. A person ceases to be a safety representative when he ceases to be employed at the workplace (unless he was appointed to more than one workplace and retains his appointment at one of those workplaces). The representative may, of course resign. The representative should "so far as is reasonably practicable" have two years' employment with the employer or at least two years' experience in similar employment. In some circumstances it will not be reasonably practicable to comply with this latter rule. These include newly-established enterprises, work of short duration or work with a high labour turnover, of which the two main examples are the construction and agricultural sectors which, to date, have amongst the lowest levels of representation (see Table 2, p. 200 below). As we have seen, section 2(4) of the Act envisages that representatives will be appointed "from amongst the employees," but the Regulations modify this in relation to employees in the group or groups the safety representatives represent who are members of the British Actors' Equity Association or of the Musicians' Union. Such representatives need not be employees and, needless to say, are not entitled to time off with pay from the employer of the employees whom they represent.

Once the safety representatives are appointed, it is the duty of every employer to consult them with a view to the making and maintenance of arrangements which will enable him and his employees to co-operate effectively in promoting and developing measures to ensure the health and safety at work of the employees, and in checking the effectiveness of such measures. As we shall see, this rather generalised duty in section 2(6) is amplified in the Regulations, Code of Practice and Guidance Notes. It is possible that disputes concerning the rights of trade unions or the rights of safety representatives could come to assume an industrial relations dimension and this no doubt goes to explain, in part at least, the reluctance of the enforcing authorities to use their statutory powers in such disputes which are best left to be settled in accordance with agreed procedure arrangements with ACAS as an ultimate "back-stop." HSC has been at pains to explain that the new legal framework should not be taken to restrict the freedom of employers and trade unions to make arrangements suitable to the circumstances of each undertaking. In some cases, the parties may wish to continue existing agreed arrangements or draw up alternative arrangements for joint consultation on health and safety at work, subject however to the right of the recognised trade union at any time to invoke the statutory rights conferred by the Act and the regulations.

HSC expect that a recognised trade union will normally appoint a safety representative to represent a group or groups of workers of a class for which the union has negotiating rights—a consideration which should not, however, inhibit the representative from raising general matters affecting the health and safety of employees as a whole.[15] In certain cases, such as a small employer with several unions represented in the workplace there is nothing to stop one representative from representing more than one group or groups of employees.[16]

There has been an interesting difference of opinion amongst trade unions as to the best persons to appoint as safety representatives. The Amalgamated Union of Engineering Workers announced its intention in its evidence to the Bullock Committtee on Industrial Democracy to restrict the appointment of safety representatives to shop stewards, a policy followed by the Transport and General Workers' Union and the National Union of Public Employees. On the other hand, at least one union, the Confederation of Health Service Employees, has recommended that safety representatives should be appointed from members who are not shop stewards with the rider that the appointees should represent a discipline, for example porters, rather than serve a geographical constituency such as a ward.

There are several advantages in appointing shop stewards as safety representatives. First, they can build on their pre-existing relationships with their employer, their own members, their union full-time officials and with other union officials. The advantage to the employer is that it may reduce the amount of time-off required and the number of people with whom he needs

[15] Guidance Notes.

[16] It will be seen that representation is based on the workplace and not the company.

to consult or negotiate. The disadvantages are principally that safety considerations may dictate that a person with a specialised knowledge should be appointed and that if a shop steward is appointed disagreements in other areas might spill over into health and safety at work. Because of the difficulties in co-ordinating the views of a large number of safety representatives in Britain's largest firms it has become necessary to have in those firms a structural hierarchy of representatives. Examples are to be found at Ford of Britain, I.C.I., British Aerospace and the Post Office. At Ford's in Dagenham, only eighteen first-tier representatives have been appointed from amongst the nineteen trade unions. Whatever system is used, it should be representation based on the *workplace* to capitalise on the knowledge and experience of those who work there and are familiar with its risks. The "workplace" means not only the place where a group or groups of employees are likely to work but also a place which they are likely to frequent in the course of their employment or incidentally to it. Flexibility can be achieved by allowing one safety representative to serve two establishments or, alternatively, by dividing one establishment into a number of separate workplaces.

A question of numbers

One of the fears held by some was that by giving trade unions a free hand in the matter of appointing safety representatives, too many representatives might be appointed consequently placing an intolerable burden upon employers. Because circumstances differ so widely between different enterprises it was found to be impossible to devise a legal formula which could be applied sensibly across industry. The Commission therefore set out its detailed advice on the number of representatives to be appointed in the Guidance Notes which accompanied the Regulations and the Code of Practice. In addition to recommending that trade unions should appoint in relation to groups of workers for which they enjoyed negotiating rights, the Guidance Notes advised that they should bear in mind:

(a) the total numbers employed;
(b) the variety of different occupations;
(c) the size of the workplace and the variety of workplace locations;
(d) the operation of shift systems;
(e) the type of work activity and the degree and character of the inherent dangers.

The Commission envisages a flexible approach both in relation to the group or groups of employees whom the representative will represent and to the numbers of representatives which might be appropriate in particular circumstances of which the following are examples:

(a) Workplaces with rapidly changing situations and conditions as the work develops and where there might be rapid changes in the level of manpower, *e.g.* building and construction sites, shipbuilding and shiprepairing, and docks.

(b) Workplaces from which the majority of employees go out to their actual place of work and subsequently report back, *e.g.* goods and freight depots, builders' yards, service depots of all kinds.

(c) Workplaces where there is a wide variety of different work activities going on within a particular location.

(d) Workplaces with a specially high process risk, *e.g.* construction sites at particular stages—demolition, excavations, steel erection, etc., and some chemical works and research establishments.

(e) Workplaces where the majority of employees are employed in low risk activities, but where one or two processes or activities or items of plant have special risks connected with them.

In addition to these examples in the Guidance Notes, many employers' organisations and trade unions have issued their own guidelines to members.[17] By 1981, one hundred and thirty thousand safety representatives had been appointed whilst some seventy thousand have received training from the T.U.C.[18] A recent survey, covering the total AUEW Engineering Section membership of 1,325,056 workers in 2436 establishments, showed that 12,080 safety representatives had been appointed of whom 8,635 were also shop stewards.[19] If a particular employer considers that too many safety representatives have been, or are about to be, appointed, two remedies are open to him. He may seek to resolve the matter by negotiation through normal industrial relations machinery, or he may refuse to grant time-off which is requested; in which event the safety representative may decide to present a complaint to an industrial tribunal under the Regulations (reg.11) on which the tribunal will have to decide.

In a survey carried out by the HSE,[20] using a questionnaire, in all premises routinely inspected by H.M. Factory Inspecto-

[17] See for example, the advice given by the Federation of Civil Engineering Contractors and the Chartered Institute of Builders.

[18] *Health and Safety at Work*, October 1981, p. 27.

[19] Reported in *Heath and Safety Information Bulletin*, January 1981, p. 13.

[20] See "Safely Appointed," DE Gazette, vol. 89, no. 2, February 1981, pp. 55–58. On Safety Committees, see below.

rate, H.M. Mines and Quarries Inspectorate and H.M. Agri-cultural Inspectorate during the period October 1–October 26 1979, it was found that only 17 per cent. of all workplaces had safety representatives, although 92 per cent. of all firms employing over 500 employees had safety representatives. The object of the survey was to ascertain the number of safety representatives appointed and the number of safety committees established, to find out how many workplaces were covered and how many safety representatives acted for more than one trade union, and, finally, to ascertain whether the appointment of safety representatives may have affected the number of com-plaints received by Inspectorates concerning workplaces. The results of the survey can be seen in Tables 1 and 2, below.

Table 2, on page 200, indicates that approximately one fifth of the total number of safety representatives acted for more than one union, whether by formal appointment or informal arrange-ment, an indication which reveals a degree of flexibility on the part of trade unions. It can also been seen that a relatively low level of recognition in small firms has led to the low level of appointments of safety representatives in those firms. It might have been preferable to follow T.U.C. advice on the HSC Consultative Document and to have allowed independent trade unions nominated to Wages Councils[21] and similar bodies to appoint safety representatives at workplaces where they have members.

Functions of safety representatives

Section 2(4) provides that the appointed safety representative shall represent the employees in consultations with the em-ployer under section 2(6) and shall have such other functions as may be prescribed. These other functions have been set out in reg. 4 of the Regulations as follows:—

(a) to investigate potential hazards and dangerous occurr-ences at the workplace (whether or not they are drawn to his attention by the employees he represents) and to examine the causes of accidents at the workplace;

(b) to investigate complaints by an employee he represents relating to that employee's health, safety or welfare at work;

(c) to make representations to the employer on matters arising out of sub-paragraphs (a) and (b) above;

(d) to make representations to the employer on general

[21] See para. 1(2) of Schedule 2 to Wages Councils Act 1979.

Table 1 Safety representatives in workplaces classified by size

Size band (number of employees)	Number of workplaces surveyed	Number of workplaces with safety representatives appointed under regulations		Average number of safety representatives in workplaces where safety representatives appointed	Workplaces with safety representatives where not all employees were represented*		Number of workpaces with safety representatives where one or more safety representative acted for more than one union		Number of workplaces with safety representatives where all safety representatives acted for more than one union	
		No.	Per cent		No.	Per cent*	No.	Per cent*	No.	Per cent*
(1)	(2)	(3)		(4)		(5)		(6)		(7)†
1– 10	3,758	99	3	1·2	9	9	34	34	33	33
11– 25	1,190	157	13	1·4	27	17	39	25	33	21
26– 50	670	199	30	1·9	52	27	57	29	35	18
51– 100	406	204	51	2·7	57	28	61	30	30	14
101– 150	294	207	70	4·2	67	32	71	34	35	17
251– 500	137	114	83	7·7	45	39	40	35	17	15
501–1,000	77	71	92	14·1	21	30	29	41	6	8
1,001 plus	98	90	92	37·7	27	30	42	47	16	18
All	6,630	1,141	17	6·5	305	27	373	33	205	18

*Percentage = as proportion of workplaces in column 3 (workplaces with safety representatives).
† The figures in column 7 are included in those in column 6.

Table 2 Safety representatives in workplaces classified by industry

SIC order	Number of employees in working places surveyed	Number of workplaces surveyed	Average number of employees per workplace	Percentage of workplaces with safety representatives appointed under regulations	Percentage of employees in workplaces with safety representatives
Agriculture, forestry, fishing*	5,595	1,129	5	1	15
Mining and quarrying†	13,282	314	42	47	85
Food, drink, tobacco	26,766	141	190	37	70
Coal and petroleum products	3,879	14	277	64	81
Chemical and allied industries	42,699	113	378	50	83
Metal manufacture	55,228	123	449	51	94
Mechanical engineering	27,395	273	100	22	80
Instrument engineering	1,338	31	43	7	43
Electrical engineering	21,541	107	201	33	88
Shipbuilding	5,423	29	187	21	92
Vehicles	57,780	83	696	45	81
Metal goods not elsewhere specified	16,059	356	45	18	68
Textiles	15,282	110	139	38	80
Leather, leather goods and fur	146	10	15	0	0
Clothing and footwear	8,128	118	69	12	60
Bricks, pottery, cement, glass	7,966	108	74	23	74
Timber, furniture etc.	6,724	206	33	11	67
Paper, printing and publishing	18,654	173	108	38	88
Other manufacturing industries	31,312	152	206	23	88
Construction	43,181	1,912	23	4	58
Gas, water and electricity	1,446	54	27	67	90
Transport and communication‡	7,784	42	185	38	98
Distributive trades (wholesale and retail)**	3,631	148	25	17	67
Insurance, banking, and business services§	104	3	35	0	0
Professional and scientific services¶	45,851	240	191	54	82
Miscellaneous services‖	9,944	526	19	8	54
Public administration and defence	9,630	115	84	61	74
All	486,765	6,630	73	17	79

Notes: * The sample consists mainly or entirely of agricultural premises.
† The sample includes no coal mines, since these are outside the scope of the regulations.
‡ The sample excludes premises subject to inspection by the Railway inspectorate.
** Many workplaces within this Order are subject to local authority inspection.
§ Workplaces within this Order are mostly subject to local authority inspection—hence tiny sample.
¶ This Order includes hospitals and education establishments.
|| Some workplaces within this Order are subject to local authority inspection.

matters affecting the health, safety or welfare at work of the employees at the workplace;

(e) to carry out inspections in accordance with regs. 5, 6 and 7 (below);

(f) to represent the employees he was appointed to represent in consultations at the workplace with inspectors of HSE and of any other enforcing authority;

(g) to receive information from inspectors in accordance with section 28(8) of the 1974 Act; and

(h) to attend meetings of safety committees where he attends in his capacity as a safety representative in connection with any of the above functions.

In order to fulfil their functions under section 2(4) of the Act, safety representatives are urged by the Code of Practice to—

(a) take all reasonably practicable steps to keep themselves informed of legal requirements, the particular hazards at the workplace and the measures necessary to eliminate or minimise them, and the health and safety policy of their employer and the organisation and arrangements for fulfilling that policy;

(b) encourage co-operation between the employer and the employees in promoting health and safety measures and in checking the effectiveness of those measures;

(c) to bring to the employer's notice normally in writing any unsafe or unhealthy conditions or working practices or unsatisfactory arrangements for welfare at work which come to their attention.

In cases of urgency, or in connection with minor matters, a report may be made orally. The Guidance Notes recommend ready access to management or their representatives *e.g.* a foreman or chargehand with responsibility for health and safety at work. Representatives will normally keep in touch with other representatives, including those appointed by trade unions other than their own, so that common problems can be resolved.

Inspections of the workplace. One of the important new rights given to safety representatives outside mining is the right to inspect the workplace or part of it on giving reasonable notice in writing to the employer of their intention to do so, provided that they have not carried out an inspection within the previous three months. The T.U.C. wanted safety representatives to have the right to make monthly inspections but the regulations allow inspections within the three-month period as a matter for agreement with the employer, unless there has been a substantial change in the conditions of work (*e.g.* new machinery has been introduced) or new information has been published by HSC/HSE relevant to the workplace since the last inspection, when the representatives may carry out a further inspection provided they do so "after consultation" with the employer. The employer is obliged to provide such facilities and assistance as the representatives may reasonably require (including facilities for independent investigation and private discussion with the employees) for the purpose of carrying out an inspection under the regulation, which inspection the employer or his representative is entitled to attend.

In contrast to routine inspections, reg. 6 entitles representatives to carry out an inspection of part of a workplace in which a notifiable accident or dangerous occurrence has taken place or in which a notifiable disease has been contracted, provided that if it is reasonably practicable to do so they must notify the employer or his representative of their intention to carry out such an inspection.[22] It is also provided that it must be safe for an inspection to be carried out and the interests of employees whom the representatives represent must be involved. Once again the employer is required to provide facilities and assistance and the opportunity for private discussions with employees and once again he is entitled to be present during the inspection. In *Dowsett* v. *Ford Motor Company Ltd.*[23] an industrial tribunal considered that the function of a safety representative in investigating potential hazards, dangerous occurrences and the causes of accidents allows him in a proper case to go outside his workplace if this is necessary to enable him to carry out that function. Short of saying that the right to go outside the workplace must necessarily be a matter of fact and degree, no clear rules were laid down concerning the occasions on which the right might justifiably be exercised. Thus the phrase in the regulations permitting the representative "to examine the causes of accidents at the workplace" does not on its face require that the examination itself must take place at the workplace.

The third fasciculus of rights under the rubric of "Inspection"

[22] The accident, occurrence or disease must be notifiable: see Notification of Accidents and Dangerous Occurrences Regulations 1980 S.I. 1980 No. 804 p. 217 below.
[23] London Industrial Tribunal, November 28, 1980 (C.O.I.T. 1975/235).

consists of the rights to inspect documents and to receive information as set out in regulation 7. Because this is part of the wider topic dealing generally with the obtaining and receiving of information to which Chapter 7 is devoted it is dealt with in that Chapter (see pp. 178 *et seq.*).

None of the functions given to the safety representative under the Regulations or indicated in the Guidance Notes may be construed as imposing any duty on him and no criminal proceedings may be instituted in respect of any act or omission by him in respect of the performance of functions so given or indicated.[24] The fact that the representative has agreed to some health and safety measure suggested by his employer, which turns out to be dangerous or unhealthy, does not expose him to criminal prosecution. Nevertheless, he may be liable under sections 7 and 8 but not *qua* safety representative discharging his functions.

Time off

The regulations add a further right to time off to the rights to time off work introduced by employment protection legislation. An employer is obliged by regulation 4(2) to permit a safety representative to take such time off with pay during the employee's working hours as shall be necessary for the purposes of:

(a) performing his functions under section 2(4) and his functions under the regulations (see above);
(b) undergoing such training in aspects of those functions as may be reasonable in all the circumstances having regard to any relevant provisions of the Code of Practice on Time off for Training.

It will be seen that a double right is involved—the right to time off and the right to pay during time off. The method of calculating pay is described in the Schedule to the regulations.

The safety representative can present a complaint to an industrial tribunal that his employer has not given him time-off or pay in accordance with the Regulations provided that the complaint is brought within three months of the employer's omission or such further period as the tribunal considers to be reasonable where it is satisfied that it was not reasonably practicable to present the complaint within the three months. If the tribunal finds that a complaint relating to time-off is well-founded, it will make a declaration to that effect and may award such compensation as it considers to be just and equitable

[24] Rider to reg. 4(1) and Guidance Note 11.

having regard to the employer's default in failing to allow time-off and to any loss sustained by the employee by reason of the matters in the complaint. So far as the complaint relates to a failure to pay the representative during time-off, the tribunal may award him such amount as it finds to be due to him.

So far as training of safety representatives is concerned, the T.U.C. regards this as being very much its own concern. The trade union role is recognised in paragraph 27 of the Guidance Notes which state—"The recognised trade unions responsible for appointing safety representatives will make their own arrangements for the information and guidance of their appointed safety representatives as to how they will carry out their functions." In a leading case,[25] the Employment Appeal Tribunal held that the question whether time-off is necessary to undergo such training as is reasonable in all the circumstances should be decided by interpreting the regulations and not by mere reliance on the Code of Practice on Time-Off or on the Guidance Notes which, although of relevance, are not decisive of the issue.

The Code of Practice prepared by HSC[26] on Time-off for the training of Safety Representatives, deals with time-off with pay "for the purpose of undergoing training approved by the TUC or independent unions," although it also refers to training by employers in the technical hazards of the workplace, on precautions and safe systems of work and on organisation and arrangements for health and safety. The argument that every training course must be approved by the T.U.C. or the unions was rejected although the recommendations of the Code might well weigh heavily with a tribunal in the circumstances of a particular case.

In that case, White was employed by Pressed Steel Fisher, a Swindon subsidiary of BL, which employed some 5,000 workers. The T.U.C. had organised a ten-day course in Swindon which White wished to attend. To enable him to do so would have involved his firm in releasing him for one day each week over a ten-week period. The firm had in fact organised on its own premises an eight-day course which was staffed by its own safety officer and wanted White to attend that course.[27] White's union, the Transport and General Workers' Union, argued that the course organised by White's firm was unsatisfactory in that it did not contain a trade union or representational element, an

[25] *White* v. *Pressed Steel Fisher* [1980] I.R.L.R. 176. See also the cases discussed by J. McIlroy, "The Education and Training of Safety Representatives," *Health and Safety at Work*, Vol. 3, no. 3, November 1980, pp. 103–4.

[26] This Code should not be confused with the ACAS Code on Time Off which was issued under employment protection legislation.

[27] Employers have a duty to provide information, instruction and training . . . as is necessary to ensure, so far as is reasonably practicable, the health and safety at work of their employees (see s.2(2)(*c*) of the 1974 Act).

argument to which the firm responded by asking the union to provide such an element. The union declined to accede to this request, insisting that the official should attend the T.U.C. course. The tribunal concluded that the firm's course was adequate save that it did not contain a trade union element. The Employment Appeal Tribunal upheld the decision of the tribunal, but remitted the case to the tribunal for it to consider whether or not it was necessary for the employee to travel to Swindon for the trade union or representational aspect only of the training. The case underlines the need for agreement between employers and unions on acceptable training courses, bearing in mind that unions are not obliged to co-operate in the framing of courses by employers. It is obviously undesirable to use Solomon's solution and have divided responsibility for courses.

A comprehensive survey recently carried out by the Amalgamated Union of Engineering Workers (Engineering Section).[28] a trade union with a membership of 1,325,056 spread over 2436 establishments in 28 different industries, showed that of the 12,080 safety representatives appointed by the union more than 70 per cent. were shop stewards. 76 per cent. of the AUEW safety representatives enjoyed formal agreements for time-off for training and regular inspections were carried out in 83 per cent. of the establishments. It was found that time-off was permitted, either formally or informally, for some 94 per cent. of appointed safety representatives.

Safety Committees

Every employer, if requested in writing to do so by at least two safety representatives, must establish a safety committee within three months of the request in accordance with the following provisions:

(a) [the employer] [must] consult with the safety representatives who made the request and with the representatives of recognised trade unions whose members work in any workplace in respect of which he proposes that the committee should function.

(b) [the employer] [must] post a notice stating the composition of the committee and the workplace or workplaces to be covered by it in a place where it may be easily read by the employees.[29]

[28] *op. cit.* p. 197.

[29] See section 2(7) of the Health and Safety at Work etc. Act 1974 and Regulation 9, Safety Representatives and Safety Committees Regulations 1977 and accompanying Guidance Notes.

Safety committees should be concerned with all relevant aspects of health, safety and welfare of persons at work in relation to the working environment.

It is suggested that safety committees should have agreed objectives or terms of reference which should include the promotion of co-operation between employers and employees in instigating, developing and carrying out measures to ensure the health and safety at work of the employees.

Under section 2(7) of the 1974 Act, safety committees have the function of keeping under review the measures taken to ensure the health and safety of the employees. In paragraph 7 of the Guidance Notes to SRSCR 1977 it is suggested that the functions of a safety committee could include:

(a) The study of accident and notifiable diseases, statistics and trends, so that reports can be made to management on unsafe and unhealthy conditions and practices together with recommendations for corrective action.

(b) Examinations of safety audit reports on a similar basis.

(c) Consideration of reports and factual information provided by inspectors of the enforcing authority appointed under the Health and Safety at Work etc. Act 1974.

(d) Consideration of reports which safety representatives may wish to submit.

(e) Assistance in the development of works safety rules and safe systems of work.

(f) A watch on the effectiveness of the safety content of employee training.

(g) A watch on the adequacy of safety and health communication and publicity in the workplace.

(h) The provision of a link with the appropriate inspectorates of the enforcing authority.

None of these functions should supplant management's responsibility for ensuring the health and safety of its employees. The work of a safety committee supplements and augments these arrangements but should not serve as a substitute for them.

Ideally the safety committee should be as small as possible and compatible with adequate representation. The number of management representatives should not exceed the number of employees' representatives. Management representatives should include line managers, supervisors, works engineers and personnel managers. Company doctors, industrial hygienists and safety officers should be ex officio members. Whilst it is clearly preferable that safety representatives should be members of the safety committee, there is no legal requirement to so include them.

Tables 3 and 4, on pages 208 and 209 below, which are taken from the HSE survey and reported in the Department of Employment Gazette, give details of safety committees in relation to the size of workplaces with safety representatives and to the industry of such workplaces.[30]

Safety committees of one sort and another were not an innovation brought in by the regulations in that a majority of firms with employees in the 251–500 range, or larger, already had a safety committee before the regulations took effect. The survey (from which the Tables are taken) looked only at those workplaces which had safety representatives. It was then established how many of those workplaces had a safety committee, distinguishing those which had been set up or altered significantly in constitution or structure as a result of regulations. It can be seen, first of all, that the majority of workplaces with safety representatives also had a safety committee (some had two committees) and, secondly, that in the first year of operation of the regulations the percentage of establishments which had safety committees rose from 50 per cent. to 81 per cent.

Interestingly, of 1141 workplaces which had safety representatives, 112 were the subject of complaints received by the Inspectorate in the first year in which the regulations were in operation, compared with 102 in the previous year. The number of complaints during that first year rose from 456 to 557. Finally, in a questionnaire-based study conducted by Mr. P. B. Beaumont, a lecturer in industrial relations in the University of Glasgow, some of the aspects of the relationship between union-appointed safety representatives and their workforce constituencies were investigated.[31] The results of the survey were obtained from an analysis of the returns of the questionnaires from 162 safety representatives attending health and safety training courses at a College of Commerce in the West of Scotland during the first half of 1979. It was found that the safety representatives were long-serving employees and that 54 per cent. were both safety representatives and shop stewards. Considerable importance was attached by the representatives to the communication of the complaints of employees to management, particularly so for those who worked in large firms in which employees were concerned about health and safety matters and the management was receptive to health and safety complaints.

[30] Department of Employment Gazette, vol. 89, no. 2, February 1981, p. 57: on safety representatives, see *ibid.*, p. 197, above.

[31] *The nature of the relationship between safety representatives and their workforce constituencies*: Industrial Relations Journal 1980, vol. 12, pp. 53–60.

Table 3 Safety committees in workplaces with safety representatives, classified by size

Size band (number of employees)	Number of workplaces*	Number of workplaces* with a safety committee		Number of workplaces* with more than one safety committee		Greatest number of safety committees in one workplace*	Number of workplaces* with safety committees in tiers		Number of workplaces* whose safety committee resulted from regulations		Number of workplaces* whose safety committee was altered significantly because of regulations	
		No.	Per cent	No.	Per cent		No.	Per cent	No.	Per cent†	No.	Per cent†
(1)	(2)		(3)		(4)	(5)		(6)		(7)		(8)
1– 10	99	82	83	2	2	2	2	2	49	60	15	18
11– 25	157	105	60	1	0·6	3	1	0·6	60	57	11	10
26– 50	199	139	70	4	2	2	3	1·5	71	51	18	13
51– 100	204	161	79	10	5	6	7	3	66	41	19	18
101– 250	207	190	92	14	7	12	8	4	67	35	39	21
251– 500	114	104	91	20	18	11	14	12	27	26	27	26
501–1,000	71	63	89	30	42	15	25	35	9	14	18	29
1,001 plus	90	85	94	53	59	45	41	46	14	16	29	34
All	1,141	929	81	134	12	45	101	9	363	39	186	20

* These are workplaces where safety representatives had been appointed, i.e. those in table 1 column 3.
† Percentage = proportion of workplaces in column 3.

Table 4 Safety committees in workplaces with safety representatives, classified by industry

SIC order	Percentage of workplaces* which had a safety committee(s)	Percentage of workplaces* with a safety committee(s), where the committee resulted from the regulations	Percentage of workplaces* with a safety committee(s), where the committee was altered as result of the regulations
Coal and petroleum products	100‖	22	11
Shipbuilding	100†	17	33
Instrument engineering	100‡	0	50
Gas, electricity and water	97	54	9
Chemicals and allied industries	97	13	38
Mines and quarries	96	51	12
Distributive trades	95	35	30
Vehicles	95	29	17
Food, drink and tobacco	94	29	29
Transport and communication	94	33	6
Other manufacturing industries	94	30	24
Bricks, pottery, glass, cement	92	30	26
Metal manufacture	90	19	23
Textiles	90	18	21
Metal goods nes	88	39	19
Electrical engineering	86	40	23
Mechanical engineering	82	32	18
Miscellaneous services	77	48	23
Paper, printing and publishing	75	29	27
Clothing and footwear	71	50	20
Agriculture, forestry and fishing	70**	70	30
Construction	69	59	14
Timber and furniture	65	53	27
Public administration and defence	64	56	24
Professional and scientific services	48	58	24
Overall	81	39	20

Notes: * These are workplaces where safety representatives had been appointed, that is those in table 2 column 4.
† A small sample of six establishments.
‡ A small sample of two establishments.
** A small sample of 10 establishments.
‖ A small sample of nine establishments.

Enforcement by inspectors

Although the system of safety representatives should be self-regulating with little need for enforcement of the law, there are three occasions on which inspectors have been advised by the Health and Safety Commission to intervene.

First, where an employer has failed to recognise the appointment of a safety representative by a recognised trade union (*cf.* a claim from the employer that the appointment has not been made by an *independent, recognised* trade union, which is ultimately a matter for the Advisory Conciliation and Arbitration Service (ACAS), or, a complaint from the safety representative to an industrial tribunal in which the employer's claim can be settled collaterally).

Second, where an employer has failed to provide information and facilities in relation to which an improvement notice could be served.

Third, where an employer has failed to discuss the establishment of a safety committee with the union following a request from two safety representatives appointed by the independent, recognised trade union—a failure which might lead to a prosecution under section 2 of the Act and the regulations.

Conclusion

There is no doubt that the system of safety representatives and safety committees represents a major step forward in the provision of a proper infrastructure for consultation between employer and employees on health and safety at work.[32] Whilst "worker participation" was a slogan which produced little legal change in the Seventies and early Eighties, worker participation in health and safety became a legal duty and, to a large extent, an increasing social reality in British industry. The right to be consulted in matters affecting his health, safety and welfare reflects not only the legitimate rights of the employee but his dignity as an individual and recognition of the contribution which he can make in this area.

At the same time there is scope for improvement. It is unfortunate that the new system came into operation at a time of economic uncertainty. Inspectors report that the recession of the early Eighties has affected the safety representative's effectiveness. Concern for job security has tended to replace concern for working conditions. The system cannot work unless

[32] See the French experience and the rules governing the Comités d'hygiene et de sécurité. These rules set out in detail how the Committee should function and they are enforceable by the Labour Inspectorate.

it is integrated into the managerial structure of the firm from the Board to the first-line supervisor such as the foreman or chargehand. Unless the representative can raise matters effectively with management the system runs the risk of becoming a dialectical forum for the exchange of views which no one takes seriously.

The system is essentially a consultative system in that management is not obliged to negotiate health and safety with his safety representatives. However, as was indicated in Chapter 1, the distinction between consultation and negotiation, however conceptually and semantically valid it may be, is likely to become blurred in practice. The T.U.C. itself supports negotiation believing that there is room for compromise, although it is recognised that this can only take place above "the floor" of statutory standards. Where there is a reluctance to negotiate, this, in the eyes of the Health and Safety Executive, seems to be affected at least in part by five factors, *viz.*—

> "by geography and by the social traditions of a particular area; by the standards of a dominant majority; by the size of the place of work; by the extent of training received by safety representatives and by the extent to which employers are prepared to train management to resolve safety issues with the representatives."[33]

When we say that the system is consultative we are referring to the statutory system just described. There is, of course, nothing to stop employers and trade unions from including health and safety at work within the ambit of collective bargaining and this frequently happens. In one case, a trade union has obtained the right to stop further work by means of a term to that effect in a collective agreement which describes the circumstances in which such an extreme "right" arises. The new system is not likely to become a major feature of those occupations categorised as "low-risk," but, even then, it is noticeable that technological changes can produce new problems.

[33] *Manufacturing and Service Industries, 1977*; H.M.S.O. 1978, at para. 42.

CHAPTER 9

SAFETY POLICIES

In its analysis of the old system, the Robens Committee found two very broad elements: legal regulation and supervision by the State and industrial self-regulation and self-help. At the level of the individual firm, the Committee recommended three prerequisites for progress towards a more effectively self-regulating system. First, there must be awareness of the importance of safety and health at work. Second, responsibilities, legal and otherwise, must be clearly defined. Third, the nature of the problems must be methodically assessed and those assessments translated into practical objectives and courses of action.[34]

The 1974 Act, supported by the existing statutory provisions and health and safety regulations made under the Act, provides the legislative framework within which self-regulation can work. This framework is filled out, as we have seen, by tiers of quasi-legislative and voluntary standards; but the hope is that the main thrust will come from safety policies generated in the firm and translated into practice by employers and employees, or, where applicable, persons other than employers, *e.g.* the self-employed, manufacturers and the like. The Act as amended gives a fillip in the direction of self-regulation by requiring employers with five employees or more to "prepare and as often as may be appropriate revise a written statement of his general policy with respect to the health and safety at work of his employees and the organisation and arrangements for the time being in force for carrying out that policy, and to bring the statement and any revision of it to the notice of all his employees"[35]

In a review drawn from the work and experience of the Accident Prevention Advisory Unit of H.M. Factory Inspectorate,[36] great importance is attached to organisation for health and safety and the rôle of management in that connection. It is pointed out that, whilst the obligation to have a written safety and health policy is a legal requirement, the

[34] Cmnd. 5034, Chapter 2, pp.14 *et seq.*

[35] Section 2(3) of the Health and Safety at Work etc. Act 1974, as amended by The Employers' Health and Safety Policy Statements (Exception) Regulations 1975 (S.I. 1975 No. 1584) (see above, p.83).

[36] HSE, *Effective Policies for Health and Safety* (1980), H.M.S.O.

policy itself must not be a "legal straitjacket." The CBI and the TUC have jointly agreed that "a clear written statement of a company's safety policy is the essential foundation for any effective safety organisation." The APAU stresses that the policy should reflect "the uniqueness and the special needs of the company for whom it is written," saying that the "document cannot be bought or borrowed, nor can it be written by outside consultants or inspectors." The Unit identifies the reasons for unsuccessful policies and the hallmarks of successful policies. The latter include the elevation of safety and health to importance equal to that of production, sales, costs and similar matters; management measures to see that the necessary precautions and steps are taken; inculcation of employees in their duties with regard to safety and health; identification of the main board or director with prime responsibility in this connection; and the dating of the policy statement to ensure that it is periodically revised.

Writing a policy

Section 2(3) of the 1974 Act demands four things—
(i) Preparation and revision of a written statements;
(ii) The Organization;
(iii) The Arrangements in force; and
(iv) Publicising of the policy.

Policy: Preparation and Revision of a written statement.

The Policy should demonstrate the high degree of importance accorded to safe working practice by the Company. It should be unique, reflect the special needs of the company and carry the authority of the boardroom.

The Policy should state its main objective and:

(a) specify that safety and health are management responsibilities ranking equally with responsibilities for production, sales, cost and similar matter;
(b) indicate that it is the duty of management to see that everything reasonably practicable should be done to prevent personal injury in the processes of production and in the design, construction and operation of all plant, machinery and equipment and to maintain a safe and healthy place of work;
(c) indicate that it is the duty of all employees to act responsibly and to do everything they can to prevent injury to themselves and fellow-workers. Although the policy is a management responsibility, it will rely heavily on the co-operation of those who actually produce the goods and take the risk;

(d) should identify the main board or managing board director/directors with prime responsibility for health and safety in order to make the commitment of the board precise and provide points of reference for any manager who is faced with a conflict between the demands of safety and the demands of production;

(e) should be dated as a means of ensuring that it is periodically revised in the light of current conditions and be signed by the Chairman, Managing Director, or whoever speaks for the organisation at the highest level and with the most authority on all matters of general concern;

(f) should state how and by whom its operation will be monitored.

Organisation

The ultimate responsibility for health and safety rests with the employer and senior management but in practice duties are delegated. The shape of the organisation for health and safety should be a reflection of the firm's overall management structure. In brief, the policy should show the following:

(a) An unbroken and logical delegation of duties throughout the management structure.

(b) The identification of key personnel who are responsible for ensuring that safe working arrangements are implemented and maintained.

(c) The definition of the rôles of functional and line management.

(d) Adequate support by all relevant functional management, *e.g.* safety officer, works engineers, designers, chemists, etc.

(e) Arrangements for the monitoring of safety performance.

(f) Means by which managerial and supervisory grades will be held accountable for their failure to manage effectively.

(g) The scale of resources available for safety and health.

Arrangements

These should include:

(a) the involvement of the safety department and line management at the planning stage and the analysis of health and safety factors in new projects;

(b) the health and safety criteria of plant, machinery and equipment being purchased by the undertaking with reference to mechanical safety and health hazards; in

addition the toxic properties of substances should be subject to evaluation at the purchasing stage;
(c) specific instructions for using machines, for maintaining safety systems and for the control of health hazards;
(d) specific training for operatives and particularly for those, like tool-setters in power press shops and timbermen in civil engineering whose activities affect the safety of their fellow workers.
(e) arrangements for medical examinations and biological monitoring;
(f) the provision of safety clothing and protective equipment;
(g) permit to work systems, particularly where more than one workforce is involved as in shipbuilding or construction;
(h) emergency and first aid procedures;
(j) procedures for visitors and contractors;
(k) relevant instructions at each level of involvement, for manager, foreman, supervisor and workmen.

Publicising the Policy

Section 2(3) requires employers to bring the statement and any revision of it to the notice of all his employees. Copies may be posted on notice boards, the policy may be distributed in booklet form to all existing employees and new starters.

Monitoring the Policy

Monitoring will assist management in determining whether the policy as stated is being effectively pursued within the organisation and secondly whether it is succeeding in its objective. Monitoring is the responsibility of senior management although some assistance may be gained from safety committees and appointed safety representatives.

In large undertakings three distinct levels of monitoring will be required, *viz.*:

(a) Main or managing board level.
(b) Division or overall production unit level.
(c) Workplaces level.

In small organisations only levels (a) and (c) will be relevant.

The Accident Prevention Advisory Unit has found that the following four areas of interest are essential to all successful monitoring schemes:

(a) The accident and ill health record of the undertaking;
(b) The standards of compliance with legal requirements and codes of practice relating to health and safety;

(c) The extent to which undertakings specify and achieve within a given timescale certain long-term objectives;

(d) The extent of compliance with the organisation and arrangements sections of the undertaking's own policy statement including in particular the systems of work developed by the company to meet its own needs.

It would be a mistake to think that the safety policy once written and communicated can then conveniently be forgotten and placed in a back drawer. In writing and publicising his safety policy, the employer in a sense is exercising a power to make sub-delegated "legislation" to which he and his employees may then be held by the enforcing authority and, if necessary, by tribunals and courts. In the Scottish case of *Armour* v. *Skeen*,[37] in which a director of roads was convicted of charges arising out of an accident involving defective scaffolding on a bridge, it was found that while he had issued no detailed safety instructions, his authority, Strathclyde Regional Council, had issued a statement of safety policy in which the responsibilities of the director (and others) included the duty to have a departmental safety policy; this duty was stressed. The director was convicted for his failure to make and publicise a departmental safety policy as was incumbent upon him under the Strathclyde Regional Council statement of safety policy. Similarly, an employer could himself be convicted by reference to standards laid down in his own policy.

[37] [1977] S.L.T. (High Court of Justiciary) 71. See under section 37(1) of the 1974 Act, p. 170 above.

CHAPTER 10

SOME RECENT DEVELOPMENTS

NOTIFICATION OF ACCIDENTS AND DANGEROUS OCCURRENCES

Notification of accidents has long been a feature of the British health and safety scene. The reporting of accidents permits the safety performance of companies and industries to be assessed and a statistical base to be assembled from which the enforcing authorities can obtain information concerning accidents, their causes and prevention. Prior to 1980, notification requirements were contained in several scattered provisions. These have now been replaced by the Notification of Accidents and Dangerous Occurrences Regulations 1980[1] which came into force on January 1, 1981. The principal purpose of these regulations is to extend the area over which there is a statutory obligation to notify accidents and dangerous occurrences.[2] Fatal and serious accidents to members of the public caused by work activities will be reportable for the first time.

Accidents involving fatal and major injuries and dangerous occurrences with the potential for inflicting major injuries must be reported promptly and directly to the enforcing authority whilst the more minor accidents will be reported to the enforcing authority by the Department of Health and Social Security after industrial injury benefit has been claimed by the employee.

These new regulations will ensure that accident details are recorded by each employer both for their own preventive programmes and so that they can be available for consultation by safety representatives[3] and by enforcing authorities.

This new accident reporting system has also reduced the administrative burden on employers, managers and the self employed. The number of forms for the reporting of accidents has been reduced to one and it is estimated that only five per cent. of notifiable accidents will be directly reportable.

Two sorts of events are covered by the new regulations. The first group consists of accidents arising out of and in connection with work which either:

[1] S.I. 1980 No. 804, subject to certain exceptions listed in Schedule 2.

[2] It was estimated that 275,000 additional notifications would result from the passage of these regulations.

[3] Parl. Deb. H.C. July 16, 1980 Col. 569 (Written Answers). See *supra*.

- result in the death of or major injury to any person; or
- result, in the case of an employee, in incapacity for work for more than three consecutive days,
- excluding the day of the accident (and Sunday or other day of rest).

This group uses the test of physical injury or economic loss, the former (other than death) denoting certain injuries specified in reg. 2(1). These include fracture of the skull, spine, pelvis or major bone in arm or leg; amputation of hand or foot; loss of sight or other injury requiring in-patient hospital treatment such as severe burns or scalds.

The second category consists of notifiable dangerous occurrences in which notification is required by reason of the occurrence itself and not by its *sequelae*. No less than forty-eight specified dangerous occurrences are listed. The first fourteen are reportable whatever type of workplace is involved. In the case of mines, a further eighteen are reportable, and in the case of quarries an additional six are reportable, whilst in the case of railways, three more categories of incident are included.

Examples of dangerous occurrences which are notifiable wherever they occur include the collapse or failure of a lift, crane, freight container, scaffold or other structure, the failure of containment in boilers, pressure vessels and pipelines, fires, electrical failures, gassings and acute ill health from isolated pathogens or infected material.

Areas where the reporting of accidents lies outside the scope of the 1980 regulations include certain matters relating to railways, merchant shipping, nuclear installations, civil aviation and agricultural operations.

A person charged with an offence under the regulations may, in addition to relying upon defences in the parent Act, seek to prove that he was unaware of the accident or occurrence leading to the commission of the offence and that he has taken all reasonable steps to have all notifiable accidents and dangerous occurrences brought to his notice.

By the Safety Representatives and Safety Committees Regulations 1977,[4] safety representatives are empowered to inspect workplaces in which a notifiable accident or dangerous occurrence has occurred, or a notifiable disease has been contracted, although their powers, whilst permitting them to take up matters with management, do not extend to the power to make a formal notification.

One of the aims of the regulations, that is to ensure that information about more minor accidents is obtained as quickly

[4] S.I. 1977 No. 500.

as possible by using DHSS procedures[5] may soon be thwarted. Notifications are only received *via* the DHSS when claims for industrial injury benefit are made. It is proposed by Clause 37(1) of the Social Security and Housing Benefits Bill,[6] which has passed its Third Reading in the House of Commons, to abolish injury benefit. Unless new amending regulations are made before this provision comes into force a valuable source of information for both enforcing authorites and safety representatives will be lost.

CLASSIFICATION, PACKAGING AND LABELLING OF DANGEROUS SUBSTANCES

The Packaging and Labelling of Dangerous Substances Regulations 1978,[7] as amended,[8] provide for the labelling of more than nine hundred listed substances. These regulations were made as a first step to meet E.E.C. Treaty obligations and the need to implement an E.E.C. Directive which itself has been subject to six amendments. The eventual objective is to provide for the labelling of all hazardous substances. The regulations came into force in two stages: on September 1, 1978, for container of two hundred litres or more, and on March 1, 1979 for other containers. Amending regulations came into force on July 1, 1981, save for regulation 2(b) which came into force on January 1, 1982.

It is provided by these regulations that no prescribed dangerous substance[9] may be supplied to any person unless it is in a container, which includes any receptacle, wrapper or other form of packaging which is designed made and fastened so as to stop the escape of contents when the container is handled normally and subject only to normal stresses and strains. The container and its fastening must be made of material not likely to be adversely affected by its contents.

The container must show the following particulars[10]:

(a) the name of the substance, being one of the names by which it is described in Schedule 1;

(b) the name and address of the manufacturer, the importer, the wholesaler or the supplier of the substance;

[5] See p. 217 above.
[6] H.C. Bill [80]: 1981–82 Session.
[7] S.I. 1978 No. 209.
[8] Packaging and Labelling of Dangerous Substances (Amendment) Regulations 1981 (S.I. 1981 No. 792).
[9] *i.e.* a substance specified in Schedule 1, but does not include preparations which include or contain such a substance.
[10] See Regulation 5, Packaging and Labelling of Dangerous Substances Regulations 1978.

(c)[11] the word or words, if any, in the relevant entry in column 2 of Schedule 1 and the symbol or symbols, if any, in Schedule 2 relating to that word or those words;
(d) the indication of the particular risk or risks referred to in the relevant entry in column 3 of Schedule 1 and set out in Schedule 3, except that this sub-paragraph does not apply to a container of 125 millilitres' capacity or less where the substance it contains is not an explosive, toxic or corrosive susbtance;
(e)[11] the indication of safety precautions referred to in the relevant entry in column 4 of Schedule 1 and set out in Schedule 4, except that this sub-paragraph does not apply to a container of 125 millilitres' capacity or less.

If a container is enclosed in one or more other containers, the outer container must be labelled as above. The particulars required on each container must be indelibly marked on the container or on a label securely fixed to the container; if the latter is preferred the entire surface must be in contact with the container. The dimensions of labels are specified and warning symbols (black on orange/yellow) must occupy one tenth of the area of the label and in no case to be less than one cm squared. Each word required by regulation 5 must be in English.

These regulations relate to the supply of prescribed dangerous substances and supply is defined in regulation 2 as—
(a) supply by way of sale, offer for sale, lease, hire or hire purchase;
(b) supply by way of commercial sample;
(c) transfer from a factory, warehouse or other establishment, whether or not in the same ownership, for further processing or use in a process or in manufacture,
whether or principal or as agent for another.

The regulations, as amended, are enforceable as if they were health and safety regulations made under section 15 of the 1974 Act, save that in any proceedings for an offence in relation to these regulations to which section 33(5) of that Act applies the daily punishment is restricted to a fine not exceeding £5 per day.

Where a prescribed dangerous substance is supplied in or from premises which are registered under section 75 of the Medicines Act 1968 (retail pharmacy business), the enforcing authority is the Pharmaceutical Society of Great Britain; but where it is supplied in or from any other shop,[12] mobile vehicle or market stall, the enforcing authority is the local weights and measures authority. In all other cases, the enforcing authority is the Health and Safety Executive. A person injured in consequence of

[11] Not applicable where the container is labelled in accordance with prescribed transport rules.
[12] Within the meaning of the Shops Act 1950.

a contravention of these regulations has grounds for a civil action for damages by virtue of section 47(2) of the 1974 Act. A defence is provided in respect of criminal proceedings that a person took all reasonable precautions and exercised all due diligence to avoid the commission of the offence.[13]

Finally, it is provided that where a container is required to be labelled in accordance with regulations and is so labelled, that labelling is deemed to satisfy the requirements of section 5 of the Petroleum (Consolidated) Act 1928,[14] and regulation 6 of the Highly Flammable Liquids and Liquefied Petroleum Gases Regulations 1972.[15]

A number of additional requirements are currently in force in relation to the labelling of dangerous substances to be "supplied." Examples includes the Poisons Rules 1978[16]; the Cosmetic Products Regulations 1978[17]; the Petroleum (Consolidation) Act 1928[18]; the Highly Flammable Liquids and Liquefied Petroleum Gases Regulations 1972[19]; and, the non-statutory Pesticides Safety Precautions Scheme.

New proposals have now been made by the Health and Safety Commission for comprehensive regulations on the classification, packaging and labelling of dangerous substances. These proposals are intended to rationalise, up-date and extend the existing legislation by means of a complete revision of the Packaging and Labelling of Dangerous Substances Regulations 1978. They will, in turn, implement in whole or in part, Directive 79/831/EEC on dangerous substances, Directives 73/173/EEC and 80/781/EEC on solvents, Directive 77/728/EEC on paints and related products, Directive 78/631/EEC on pesticides and Directive 78/319/EEC on toxic and dangerous wastes.[20]

CLASSIFICATION AND LABELLING FOR CONVEYANCE BY ROAD, RAIL, SEA AND AIR

The Dangerous Substances (Conveyance by Road in Road Tankers and Tank Containers) Regulations 1981,[21] save for four provisions,[22] came into force on January 1, 1982. These

[13] Regulation 7(4), S.I. 1978 No. 209, *op. cit.*
[14] 1928, c.32.
[15] S.I. 1972, No. 917.
[16] S.I. 1978 No. 1.
[17] S.I. 1978 No. 1354.
[18] 18 and 19 Geo. 5, c.32.
[19] S.I. 1972 No. 917: these regulations relate essentially to storage.
[20] Partial implementation of this Directive has already been achieved by the Control of Pollution (Special Waste) Regulations 1980 (S.I. 1980 No. 1709).
[21] S.I. 1981 No. 1059.
[22] Regulations 8, 10 and 21 concerning information and driver training will come into force on January 1, 1983 and regulation 7 concerning regular examination and testing will come into force on January 1, 1984.

regulations impose requirements for notices to be displayed on road tankers and tank containers which are being used for the conveyance of "dangerous substances." "Dangerous substances" are substances which are specified in the list entitled "Approved Substance Identification Numbers, Emergency Action Codes and Classifications for Dangerous Substances Conveyed in Road Tankers and Tank Containers," the approved list published by the Health and Safety Commission,[23] and any other substances which have characteristic properties specified in Schedule 1. Associated with the regulations is an approved code of practice on the classification of dangerous substances, which enables a substance not on the 'approved list' or a mixture, to be identified as dangerous or otherwise, and the correct warning sign to be allocated.[24]

The regulations also impose duties on the operators of road tankers and tank containers used for the conveyance of dangerous substances to ensure that any vehicle used for such conveyance is fit for the purpose and that the tanks of road tankers and tank containers are regularly examined and tested. Two approved codes of practice are being prepared in connection with this requirement, one dealing with the construction and testing of tankers and tank containers for all applications and the other with operational matters, including loading and unloading, mixed loading and safety equipment.

The regulations further provide that certain dangerous substances may not be conveyed in road tankers and tank containers and that where dangerous substances may be conveyed that precautions be taken against fire and explosion. The regulations prohibit the overfilling of any tank which contains a dangerous substance and in certain cases require vehicles to be supervised when not being driven.

It is additionally provided that an operator shall not convey a dangerous substance in a road tanker or a tank container unless he has obtained information from the consignor to enable him to be aware of the risks created by that substance, and that an operator shall ensure that the driver of a vehicle carrying a dangerous substance is given information to enable him to know the identity of the substance and its risks. The regulations provide that the driver shall ensure that such information is kept in the cab of the vehicle. Further, the operator of a vehicle used for the conveyance of a dangerous substance must ensure that the driver has received adequate instruction and training.

[23] This list is based on a recommendation of the U.N. Committee of Experts on the Transport of Dangerous Goods. (The Approved List), HS (R) 10. H.M.S.O.

[24] Approved Code of Practice Classification of Dangerous Substances for Conveyance in Road Tankers and Tank Containers.

A general guidance note on the regulations and guidance for enforcing authorities has now been published. A consultative document is to be issued on the precautions to be taken in the conveyance by road of dangerous substances other than those in road tankers or tank containers with a view to making the Dangerous Substances (Conveyance by Road of Packaged Goods) Regulations.

Further requirements are made by the Explosives Act 1875[25] and subordinate legislation the Radioactive Substances (Carriage by Road) (Great Britain) Regulations 1974[26] and by the Poisons Rules 1978.[27]

As a signatory to the European Agreement concerning the International Carriage of Dangerous Goods by Road, the United Kingdom is obliged to allow road vehicles carrying certain[28] dangerous goods on international journeys to carry such goods in the United Kingdom providing that the provisions of the Agreement are complied with. Certain road services operated by British Rail connecting with railway services may be subject to the provisions of international regulations concerning the carriage of dangerous goods by rail. Packaged goods carried by sea or by air need to be classified, packaged and labelled in accordance with the International Maritime Dangerous Goods Code or the regulations of the International Air Transport Association.

CLASSIFICATION AND LABELLING OF EXPLOSIVE ARTICLES AND SUBSTANCES

The regulation of the non-military use of explosives in the United Kingdom is governed by two principal acts, The Explosives Act 1875[29] and the Explosives Act 1923.[30]

Explosive articles and substances, other than military explosives , are currently classified under the Explosives Act 1875 and Orders in Council made thereunder. Military explosives are classified and listed by the Ministry of Defence.

The labelling and marking requirements for civil explosives are prescribed in The Packing of Explosive for Conveyance Rules,[31] Fireworks Act 1951[32] and The Merchant Shipping (Dangerous Goods) Rules.[33]

[25] 38 and 39 Vict., c.17.
[26] S.I. 1974 No. 1735.
[27] S.I. 1978 No. 1.
[28] Save for explosives, flammable liquids and gases.
[29] 38 and 39 Vict., c.17.
[30] 13 and 14 Geo. 5, c.17.
[31] S.I. 1949 No. 798.
[32] 14 and 15 Geo. 6, c.58.
[33] S.I. 1978 No. 1543.

Similar requirements for military explosives are prescribed in the Regulations for the Conveyance of Military Explosives and a Ministry of Defence manual.

The United Nations Committee of Experts on the Transport of Dangerous Goods has recently developed a scheme of classification, and this is currently being adopted by an increasing number of states and international regulatory agencies. This system has already been adopted by the Ministry of Defence in the United Kingdom and it is proposed to adopt this as an overall standard for the use of explosives in this country.

FIRST AID AT WORK

The Health and Safety (First Aid) Regulations 1981[34] which were made on June 29, 1981, came into operation on July 1, 1982. These regulations, in common with others in this field, such as the Safety Representatives and Safety Committees Regulations 1977[35] and the Control of Lead at Work Regulations 1980,[36] have been issued in conjunction with an Approved Code of Practice which takes effect on the same day as the Regulations and provides practical guidance for employers and the self-employed on how their duties might best be met. Guidance Notes have also been issued to give further advice on such matters as equipment and training.

These new regulations will extend first aid provisions to the Health and Safety at Work Act's new entrants, but rigid specifications concerning the number of first aiders and the number and content of first aid boxes have been avoided and a large number of existing provisions, particularly those made under the Factories Act and the Offices, Shops and Railway Premises Act, have been repealed. Greater emphasis is now placed on the training of first aiders and on ensuring that they will be equipped to deal with their respective workplace situations.

Perhaps the most important regulation is Regulation 3(1). It is there provided:

> "That an employer shall provide, or ensure that there are provided, such equipment and facilities as are adequate and appropriate in the circumstances for enabling first aid to be rendered to his employees if they are injured or become ill at work."

[34] S.I. 1981 No. 917.
[35] S.I. 1977 No. 500, see above.
[36] S.I.1980 No. 1248.

By virtue of regulation 3(2), an employer is further charged with the duty to provide an appropriate number of adequately qualified and trained persons to render first aid to his employees. Temporary absence of the trained first aider may be filled by an appointed person. Such a person, although not qualified or trained in first aid, should be able to ensure that an injured or ill employee is given treatment and access to appropriate equipment or facilities.

In deciding how many first aiders are required, the employer is advised in the Approved Code of Practice to bear in mind—

(a) number of employees;
(b) the nature of the undertaking;
(c) the size of the establishment and the distribution of employees;
(d) the location of the establishment and the locations to which employees go in the course of their work.

Because of the complex interaction of these categories, it is difficult to lay down precise rules to cover every circumstance; but it is surely a step forward to recognise that factors other than the type of premises and the numbers employed therein are important in determining the facilities likely to be required.

The Code recommends by way of a guide that in establishments with relatively low hazards, such as offices, shops, banks, or libraries, a first aider should not be necessary, unless 150 or more employees are at work, and that when that number is exceeded a rate of one first aider per 150 employees will probably be adequate.

In establishments with a moderate degree of hazard, such as factories, dockyards, warehouses and farms, one first aider should be present when the number of employees at work is between 50 and 150. Where there are more than 150 employees at work, there should be at least one additional first aider for every 150 or so employees.

Employers whose establishments present special hazards, such as shipbuilding, chemical industries and quarrying, should provide, instead of a first aider specified above, at least one occupational first aider. An occupational first aider has, like a first aider *simpliciter*, received a general training and has been awarded a qualification approved by the Health and Safety Executive, but, in addition, he has also received specialised instruction concerning the particular first aid requirements of his employer's undertaking.

In the circumstances described above, where a first aider is not required, an "appointed person" must be available at all times when employees are at work. An "appointed person" must be capable of taking charge of a first aid box and be available at all times to summon medical assistance.

In establishments where there is shift-working, the employer should ensure that sufficient first aiders are appointed to provide coverage for each shift and where four hundred or more employees are at work in an establishment a suitably equipped and staffed first aid room[37] should be provided, although it may be prudent to provide one where fewer people are employed if the undertaking is hazardous or if it is located some distance from medical facilities.

Employers should also ensure that every employee has quick and easy access to first aid. In a compact establishment, centralised facilities might well be suitable, whilst in a workplace with scattered centres of operation, supplementary equipment and personnel may be needed to be provided at those centres. Pursuing the same line of thought, the Code advises that where a place of work is a long distance from National Health Service accident and emergency facilities, or, if transport facilities are limited, the employer may need to provide a first aid room, even where the numbers employed would suggest that this is not necessary. It may also be necessary to provide an occupational first aider.

Small travelling first aid kits should be provided where employees work in isolated locations or where workers use potentially dangerous tools or equipment. In those circumstances in which more than one employer are working together and they wish to avoid duplication of first aid provision, the Code suggests that it might be beneficial if a written agreement were to be made, whereby one of them agrees to provide all the necessary first aid equipment, facilities and personnel. Employers who make such agreements are not absolved from ensuring that what is provided is adequate and appropriate for his employees.

Regulation 4 states:

> "An employer shall inform his employees of the arrangements that have been made in connection with provision of first aid including the location of equipment facilities and personnel."

The regulation applies notwithstanding that an agreement has been made with another employer to provide the necessary equipment, facilities and personnel. It is suggested in the Code that this information might best be given during induction

[37] For conditions relating to the operation of a first aid room see the Approved Code of Practice, paras. 26 and 27.

training and that, in addition, at least one notice, in English, might be provided in a conspicuous position in such workplace.

The self-employed person is also charged with a duty to provide, or ensure that there is provided, such equipment, if any, as is adequate and appropriate in the circumstances to enable him to render first-aid to himself while he is at work[38] (and, presumably, not unconscious). In most situations, this may simply mean a small travelling first aid box and where a number of self-employed persons work together, or a self-employed person works in premises under the control of an employer, in order to avoid duplication of facilities in agreement; an agreement may be made whereby one of them provides the equipment in which case the other must still ensure that adequate and appropriate provision is made.

[38] Regulation 5.

CHAPTER 11

THE CONTROL OF MAJOR HAZARDS

New problems

The problem from what might loosely be termed major hazards has grown steadily over the last one hundred and fifty years, but more particularly since the end of the Second World War. New materials, new processes and new industries have brought in their train new health and safety problems. The problem of the major hazard is also related to scale. If the numbers of accidents and the numbers affected by accidents are reducing, as is generally the case, it is noticeable that the consequences of individual accidents can be more severe and even catastrophic. Air safety is constantly improving, but a "jumbo jet" disaster is in itself more serious than an accident involving a small aeroplane. Thus, for sound economic reasons the numbers of installations per site have markedly grown. Whilst this development does not mean that there would be fewer accidents if, instead, there were to be a replication of smaller units, it does mean that when an accident occurs on one of these highly complex sites there are likely to be more fatalities, even though major incidents do not usually generate their full potential.[1]

The last two decades have witnessed a succession of serious accidents throughout Europe and the United States of America—the Aberfan Disaster (1966), the Flixborough Tragedy (1974), the Nuclear Incident at "Three Mile Island" (1979—Harrisburg in the U.S.A.) and the Seveso Escape in Northern Italy (1976) have led to calls for more stringent controls over industrial activities.

The failure of the common law

The redress of harm to property or personal injury to the individual is the main function of the law of tort. The law of tort still remains, despite the false prophecies of those who said that

[1] For example the fatality rate in oil processing units is thought to be a function of capacity to the power 0.5 or less. *i.e.* between $Fr = KS^{0.5}$ and $Fr = KS^{0.33}$
where Fr = Fatality rate
S = Stream capacity of oil/natural gas
K = Constant
See Marshall V.C. What are Major Hazards January 1977, on unpublished paper prepared for the Advisory Committee on Major Hazards of the Health and Safety Commission.

he Pearson Commission[2] would give it its quietus. In fairness to he law of tort it has attempted to cope with the problem of special dangers" by using the ancient and anomalous rules oncerning escapes of fire; by means of the Rule in *Rylands* v. *'letcher*[3]; and by stiffening the standard of care in common law egligence.[4] The response of the common law has been to invoke tandards of strict liability in relation to these special dangers. 'he Law Commission, in its review of the law on civil liability or dangerous things and activities, found the common law rules n these dangers to be "very complex and subject to numerous ncertainties" and stated that—

> "So far as there is any general principle, on which strict liability in the wider sense (as distinguished from liability for the fault of an independent contractor) is imposed, it seems to be as follows: *strict liability is justified in respect of certain things and certain activities which involve special dangers—i.e. a more than ordinary risk of accidents or a risk of more than ordinary damages if accidents in fact result.*"[5]
> The special danger of the Law Commission is the "ultra-hazardous activity" of the American Law Institute in its Restatement of the Law of Tort (1938), *i.e.* an activity which (a) necessarily involves a risk of serious harm to the person, land or chattels of others which cannot be eliminated by the exercise of the utmost care and (b) is not a matter of common usage. The person who carries on such an ultra-hazardous activity should be liable to those whose persons, land or chattels he should recognise as likely to be harmed by the unpreventable miscarriage of the activity for harm resulting thereto from that which makes the activity ultra-hazardous, although the utmost care is exercised to prevent the harm.

The main attempt in this country to cope with special dangers vas the Rule in *Rylands* v. *Fletcher*, stated as follows in the amous words of Blackburn J. in the Court of Exchequer Chamber—

> "We think that the true rule of law is, that the person who for his own purposes brings on his lands and collects there anything likely to do mischief if it escapes[6] must keep it in at his peril, and if he does not do so, is prima facie

[2] Royal Commission on Civil Liability and Compensation for Personal Injury (1978), Cmnd. 7054–I.
[3] (1866) L.R. 1 Ex. 265; (1868) L.R. 3 H.L. 330.
[4] See generally Winfield & Jolowicz on Tort, 11th. edition by Professor W.V.H. Rogers, particularly Chapter 16, pp. 398–432.
[5] Law.Com. No. 32. H.C. Paper No. 142.
[6] There must be an "escape": see *Read* v. *Lyons & Co. Ltd.* [1941] A.C. 156.

> answerable for all the damage which is the natura consequence of its escape."[7]

On the face of it the Rule would seem adequate to deal, if no with the prevention of the escapes or emissions of dangerou things,[8] then at least with the compensation of those who suffe loss as a necessary consequence of those escapes. That the Rul has, for all practical purposes, failed in this respect is due to th technicalities with which it became encumbered—the notion o an "escape," the need (at least initially) for the victim of th escape to have some proprietary interest in the adjoining land, the doubt concerning personal injuries, the requirement that th escape must result from "non-natural user" of the land and finally, the defences such as "Act of God" or "act of stranger. The limitations of the Rule can be seen a case such as *Dunne* v *North Western Gas Board,*[10] where gas explosions took place a manhole points in Liverpool, causing damage and physica injuries. How the gas came to be ignited was never shown although it was proved that a gas main had broken as the resul of a leak from a water main which caused the sewer beneath it t collapse, so that soil was washed away down the sewer, thereb removing support from the gas main where it broke. Liverpoo Corporation controlled both the water main and the sewer whilst the Gas Board controlled the gas main. Both were sued o a number of grounds one of which was the Rule in *Rylands* v *Fletcher*. Neither was held liable under that Rule. The Ga Board had carried out without negligence[11] the statutory dut imposed on it and could not be liable under the Rule for ga which it collected and distributed for the "common good" and no for the Gas Board's "own purposes." Sellers L.J. considered tha "in the present time the defendant's liability in that case (*i.e* *Rylands* v. *Fletcher*) could simply have been based on th defendants' failure of duty to take reasonable care."[12] Similarly Liverpool Corporation, in the absence of proved negligence, wer not liable under the Rule. It is not surprising that the learne editor of Winfield and Jolowicz finds it "a logical inference fron this and from the judgment as a whole that the Court of Appea considered the rule to have no useful function in modern times."

[7] (1866) L.R. 1 Ex.265, pp. 279–80.

[8] The remedy of the injunction, a discretionary remedy, is of limited preventive value.

[9] There are *dicta* which widen the application of the Rule to those without a proprietary interest, *e.g* the pedestrian on a road injured by a blast from an explosion in the defendant's quarry (*Miles* v *Forest Rock Granite Co. (Gloucestershire) Ltd.* (1918) 34 T.L.R. 500. The suggestion (*per* Lorc Macmillan in *Read* v. *Lyons* (above)) that personal injuries are not covered has been doubted.

[10] [1964] 2 Q.B. 806.

[11] *Geddis* v. *Proprietors of the Bann Reservoir* (1878) 3 App.Cas. 430, holding a statutory undertake liable for the negligent performance of a statutory power.

[12] S.C. at p. 831.

What started out as strict liability has very much ended up as the more relaxed duty to take reasonable care.

Apart from this standard of care which forms the essence of the tort of negligence, the law of nuisance (including public nuisance), and the law on the escape of fires have not, like the *Rylands* v. *Fletcher* rule, been able to keep the common law in tune with a rapidly changing industrial environment. If the history of strict liability in relation to "dangerous things" has not been entirely successful in coping with the new industrial risks, there remains the "safety net" of the tort of negligence. With the vast increase in governmental powers since the nineteenth century or the growth of "collectivism" as Dicey put it, the law of negligence has been extended to public authorities acting under statute. Although there can be no liability on a public authority for discharging a statutory duty, the manner of discharging that duty, or the discharge of a "permissive" statutory power, must be done without negligence. As Lord Blackburn put it—

> "For I take it, without citing cases, that it is now thoroughly established that no action will lie for doing that which the legislature has authorised, if it be done without negligence, although it does occasion damage to anyone; but an action lies for doing that which the legislature has authorized, if it be done negligently."[13]

Proving negligence has the major drawback that it places a heavy onus of proof upon the plaintiff. One has only to read the reports of the scientists on the Seveso Incident in Northern Italy or the Spanish case on contaminated cooking oil, in which the exact causes of the accidents remained shrouded in mystery, to realise that a plaintiff would have a well-nigh impossible task in bringing home negligence in such cases. In such circumstances the Courts would not assume that the accident was necessarily caused by negligence (*res ipsa loquitur*). In any case, the success of an action for negligence, like success in any legal action, depends upon the means of the defendant or the insurance cover which he has arranged to cover his negligence. Tragic as the Thalidomide Case was it would have been even more tragic had the defendants, instead of being a wealthy pharmaceutical company, been a small "back-street" pharmacy.

Piece-meal statutory modifications

Given the above limitations of the common law it is not surprising that particular problems, such as pollution of the

[13] *Geddis* v. *Proprietors of the Bann Reservoir* (1878) 3 App.Cas. 430 at pp. 455–6.

environment and the dumping of poisonous waste, have been dealt with by special statutes.[14] One of the major public concerns has been public apprehension concerning the use of nuclear energy for peaceful purposes, an apprehension which has not been allayed by the remarkable safety record of the nuclear energy industry, one or two incidents, such as the escapes at Windscale, excepted. The fear is that a major accident, involving a core melt-down in a nuclear reactor, could have catastrophic consequences not only in the short-term but also in the long-term (including possibly inter-generational consequences). The Nuclear Installations Act 1965 is based on the recognition of the need for special regulation of a special risk. Control of nuclear installations and operations is achieved under that Act in several ways.

(a) Nuclear installations, except those used by the United Kingdom Atomic Energy Authority, are confined to sites licensed for that purpose by the Health and Safety Executive.[15] Even though a nuclear site licence is in force, no person (other than the UKAEA) may use any site for (i) the treatment of irradiated matter involving the extraction of plutonium or uranium; or (ii) the treatment of uranium so as to increase the proportion of the isotope 235 contained therein, without a permit in writing for the use of the site for research or development granted by the UKAEA or a government department.[16]

(b) Section 7 of the 1965 Act imposes strict liability on the licensee in respect of certain occurrences involving nuclear matter, or the emission of ionising radiations, causing personal injury or damage to property, *i.e.* it is no defence for the licensee to say that he took all reasonable care, although he can shelter behind the defence in section 13(4)(*a*), *viz.* that the breach of duty "is attributable to hostile action in the course of any armed conflict, including armed conflict within the United Kingdom." Section 7 applies however in respect of occurrences attributable to natural disasters including exceptional and unforeseeable natural disasters, such as an earthquake. Section 11 of the Act attaches liability for injury or damage to carriers of nuclear materials within the U.K. In addition to public fears for the safety of the nuclear industry, there are fears concerning "security."

(c) Section 24 permits (but does not mandate) the Secretary of

[14] See Control of Pollution Act 1974; Gas Act 1965

[15] s.1, as amended.

[16] s.2: ss.3–5 contain further provisions on licences.

tate to appoint as inspectors to assist him in the execution of his Act (other than provisions mentioned in Schedule 1 to the 974 Act) such numbers of persons appearing to him to be ualified for the purpose, as he from time to time considers ecessary or expedient. The inspectors appointed under this ection have become known as the "Nuclear Installations Inpectorate" although this is not a legal term of art.[17] The powers f the inspectors are the same as those given to inspectors by ection 20(2) of the 1974 Act. Like the latter, sections 28 restrictions on disclosure of information), 33 (offences) and 39 prosecutions) apply to them. In effect, the Nuclear Installations nspectorate has been transferred to HSE which is, therefore, esponsible in effect not only for assessing the safety of proposed ites and the design of nuclear plant, but also for the establishment of safety requirements for the protection of employees and he public and the carrying out of inspections of nuclear nstallations.

(d) Finally, section 22 requires dangerous occurrences to be eported whether they occur on a licensed site or during the arriage of nuclear materials.

It will be seen, therefore, that the 1965 Act, as amended, urports to place regulatory controls over nuclear installations nd the transport of nuclear materials, using techniques of censing, permits, and inspection. The Act also deals with the uestion of liability for compensation to those who suffer ersonal injury or damage to property, subject to the limit of ggregate compensation mentioned in section 16.

So far as the nuclear generation of energy is concerned, the arly Magnox stations will, in the not too distant future, be nding their useful life and are currently being replaced by dvanced Gas-cooled Reactors (AGR's). At the time of writing olicy decisions are under consideration as to the scale of future eliance on nuclear power and on the desirability of using the impler and less costly Pressurised Water Reactors (PWR's). here is some concern about the safety of the latter type of eactor following the incident at "Three Mile Island" in the Jnited States of America, notwithstanding the conclusion that n unusual concatenation of human errors was mostly to blame or that incident. Future nuclear policy is currently a matter of ublic inquiry, leaving technical questions concerning the WR's, to detailed assessment which, if favourable, could bring his type of reactor into use by 1990. In the final analysis, the afety measures for nuclear generation could be so costly that, ar from being the "cheap power" as originally thought, it could

Responsibility for the issuing of permits for nuclear reprocessing and enrichment plants was not transferred to HSE unlike the licensing of nuclear reactors.

be as expensive as generation from fossil fuels or from sola energy. It is significant that at a time of public expenditure cuts the Nuclear Installations Inspectorate, far from being cut will b substantially increased,[18] notwithstanding that the Inspecto rate has in the past been provided with more staff pe operational unit than the remaining inspectorates of the Healt and Safety Executive.

The control of dangerous things is therefore a mixture o vague common law rules and special statutory intervention such as the Nuclear Installations Act 1965 discussed above. Th device of strict liability used in that statute has been used i connection with civil aviation (the Civil Aviation Act 1949) an the underground storage of gas (the Gas Act 1965). The Pearso Commission on Civil Liability and Compensation for Persona Injury[19] recognised "that there is no sharp line between thing or activities which are inherently dangerous and those whic are not. The most innocent object can be a source of danger i wrongly used." But the Commission went on to distinguis between things or operations which are of an unusuall hazardous nature (*e.g.* explosives, and flammable gases o liquids) and suggested a comprehensive statutory schem consisting of a parent Act and statutory instruments mad thereunder which would make the controller of any liste dangerous thing or activity strictly liable for death or persona injury resulting from its malfunctioning, the listing to b contained in the statutory instrument. This statutory scheme with the available defences which might be pleaded, might, i appropriate cases as specified by the statutory instrument, b backed up by compulsory third party insurance.

It should be remembered that the common law on dangerou things and the various suggestions for reform, such as those o the Law Commission and the Pearson Commission, are essen tially concerned with the question of compensation to which th prevention and regulation of hazards are subsidiary. The specia statutes, such as the Nuclear Installations Act 1965, ar "regulatory" in the sense that they seek to regulate certai activities, but they also not uncommonly include provision which deal with civil compensation to those who suffer injury t the person or damage to property. Once again we find preventio and reparation joined in a symbiotic union.

The provisions in Part I of the Health and Safety at Work etc Act 1974 are expressed to have effect with a view to—

(a) securing the health, safety and welfare of persons at work

[18] HSC Plan of Work 1981–82 and onwards (1981).
[19] Report, Cmnd. 7054–1, Chapter 31.

(b) protecting persons other than persons at work against risks to health or safety arising out of or in connection with the activities of persons at work;

(c) controlling the keeping and use of explosive or highly flammable or other dangerous substances, and generally preventing the unlawful acquisition, possession and use of such substances; and

(d) controlling the emission into the atmosphere of noxious or offensive substances from premises of any class prescribed for the purposes of this paragraph.

If further regulation is required there is, as we have seen in Chapter 4, power to make regulations which will give rise to civil liability in the absence of provision therein to the contrary. The early view that regulations would supersede the existing statutory provisions whilst still, no doubt, a long-term objective, has been to some extent overtaken by a later view according to which the regultions should at least in the short-term deal with new and pressing problems in a rapidly-changing industrial situation. One of the matters which is concerning the Health and Safety Commission is the problem of controlling major hazards. As we saw above, the spate of disasters in recent years has called for a tighter degree of control on those who create new hazards either by the use of new substances and processes or by the increased scale of traditional activities. The disaster which set in train the demand for tighter controls in this country was the disaster at Flixborough in Humberside which occurred on Saturday, June 1, 1974, at the chemical works of Nypro (U.K.) Ltd., a company jointly owned by Dutch State Mines, the National Coal Board and Fisons Ltd. At the time of the accident, the plant capacity was 50,000 tons of caprolactum per year. At 4.53 p.m. a temporary by-pass pipe failed resulting in the release of cyclohexane, a chemical having similar properties to petrol above its boiling point, at a rate faster than one ton per second from the two pipe stubs. After about forty-five seconds the vapour cloud ignited producing rapid deflagration and a powerful pressure wave as the remaining one thousand tons of cyclohexane caught fire.

Within the 24 hectare works area, 28 people were killed and 36 others were injured, whilst outside the works area, 53 people were injured and 2000 buildings were damaged. The Fire Protection Association estimated the cost of the damage at £36m.[20] The accident was the cause of great public concern and

[20] The Flixborough Disaster. Report of the Court of Inquiry HMSO 1975 (Chairman R. J. Parker, Q.C.).

it led to the setting up by the Health and Safety Commission of the Advisory Committee on Major Hazards in January 1975. The Committee was given the following terms of reference:

> "To identify types of installations (excluding nuclear installations) which have the potential to present major hazards to employees or to the public or the environment, and to advise on measures of control, appropriate to the nature and degree of hazard, over the establishment, siting, layout, design, operation, maintenance and development of such installations, as well as over all development, both industrial and non industrial, in the vicinity of such installations."

The Committee first attempted to identify those installations which could present a major threat to the safety of employees or the general public arising from explosion, the sudden release of a toxic substance, or cataclysmic fire.

In the case of flammable materials, the greatest threat arises from the sudden massive escape of volatile gases or substances which could produce large clouds of flammable, possibly explosive vapour. A number of factors, for example, wind speeds and the dilution of the cloud with air, determine the effect of such an explosion; but one important factor is the very short interval between the initial escape and the fire and explosion which, in the case of the Flixborough Disaster, was less than one minute.

With toxic materials, the sudden release of large quantities of dust or liquids, exemplified by the accident at Seveso in Northern Italy in 1976, can cause even larger numbers of fatalities and can leave lasting consequences on the landscape. Some installations pose both types of threat. Where storage facilities and industry are concentrated, as for example on Canvey Island and the neighbouring part of Thurrock, there is the possibility of a mammoth conflagration as the integrity of each plant is successively breached.

PROPOSALS OF THE ADVISORY COMMITTEE ON MAJOR HAZARDS[21]

Notification

The Committee decided that the notification of hazards was fundamental to the mechanism of control and that occupiers of

[21] The Committee issued its First Report in 1976 with the suggestion of a notification scheme which was generally favourably received. The Second Report (1979) followed the publication in 1978 of a Consultative Document on Hazardous Installations (Notification and Survey) Regulations by the HSC and, like the latter, was open for comment.

certain installations should be required to send to HSE specified details of their activities. Both existing and proposed installations should be so notified, especially where a proposal to make changes would give rise to a notifiable activity.

It was therefore recommended that new notifiable installations regulations should be made and that the following list should constitute notifiable installations.

INVENTORIES REQUIRING NOTIFICATION

Substance	Quantity
Group 1 *Toxic substances*	
Phosgene	2 tonnes
Chlorine	10 tonnes
Acrylonitrile	20 tonnes
Hydrogen cyanide	20 tonnes
Carbon disulphide	20 tonnes
Sulphur dioxide	20 tonnes
Bromine	40 tonnes
Ammonia	100 tonnes
Group 2 *Substances of extreme toxicity*	
Toxic liquids or gases likely to be lethal to man in quantities of less than one milligramme	100 grammes
Toxic solids likely to be lethal to man in quantities of less than one milligramme other than those which are and which will be maintained at ambient temperature and atmospheric pressure.	100 grammes
Group 3 *Highly reactive substances*	
Hydrogen	2 tonnes
Ethylene oxide	5 tonnes
Propylene oxide	5 tonnes
Organic peroxides	5 tonnes
Nitrocellulose compounds	50 tonnes
Ammonium nitrate	500 tonnes
Sodium chlorate[22]	500 tonnes
Liquid oxygen	1,000 tonnes
Group 4 *Other substances and processes*	
Flammable gases not specified in any other group	15 tonnes
Flammable liquids above their boiling point (at 1 bar pressure) and under pressure greater than 1.34 bar including flammable gases dissolved under pressure but not mentioned in any other category.	20 tonnes
Liquefied petroleum gases such as commercial propane and commercial butane and any mixture thereof.	30 tonnes
Liquefied flammable gases under refrigeration which have a boiling point below 0°C at 1 bar pressure and are not included in Groups 1–3.	50 tonnes
Flammable liquids of flash point less than 21°C not included in Groups 1–3.	10,000 tonnes
Compound fertilisers.	500 tonnes
Plastic foam	500 tonnes

In deciding notifiable levels, the Committee considered three main factors—

(a) General toxicity.
(b) Physical properties.
(c) Vapour pressure.

The proposed Hazardous Installations (Notification and Survey) Regulations provide that a person, who at any one time keeps, manufactures, possesses or uses a notifiable quantity of hazardous substances in any place must supply the following information in relation to his undertaking.

(1) The name and address of the person making the notification.
(2) The address and postal code of the place to be notified, or ordinance survey grid reference.

[22] See the fire and explosion which occurred on the Thames View Estate, Barking, Essex, January 21, 1980 when two tonnes of sodium chlorate caught fire and exploded. Also the fire and explosion at Braehead Container depot, Renfrew, January 4, 1977 (HMSO).

(3) The approximate area of the installation or place covered by the notification.

(4) A general description of the activities carried on, *e.g.* oil refinery, water treatment works, fertiliser store and research station.

(5) For notifiable substances or processes, a list of those which are kept or are expected to be kept in excess of the notifiable levels and the estimated maximum quantities for each or generated by or contained in each such process.

The Committee estimated that these requirements would lead to some five thousand notifications.

Hazard Surveys

The Committee felt that it was the duty of a company operating a notifiable installation to survey the hazards to which its undertaking gives rise, to identify its own problems and to set up appropriate machinery and procedures for solving them. It recommended that the company should be required to make a survey of the hazard potential of its plant and to inform the Health and Safety Executive of both the hazards which have been identified and the procedures and methods in force or which will be adapted to deal with them.

In some cases, particularly where there are the highest risks, or those involving novel or rapidly changing technologies, a more elaborate assessment might be required by the Health and Safety Executive.

The proposed Hazardous Installations (Notification and Survey) Regulations provide that in certain cases[23] a hazard survey should be carried out, the objectives of which are—

(1) To provide information and data on the nature and scale of the hazards involved.

(2) To identify the more critical features of the hazardous undertaking and of the systems used to control the hazards.

(3) To cause the operating organisation to review the events which could lead to loss of containment of the most hazardous inventories and the consequences of such loss.

(4) To consider the effects of potentially catastrophic events even if rare.

The conclusions which the Health and Safety Executive would be able to derive from such a survey would enable them to allocate their resources for surveillance more effectively and to evaluate the threat to public safety. They would also be in a

[23] Expected to affect some three hundred premises.

better position to advise local planning authorities in relation to the statutory duties imposed on them. Since the number of trained persons available both to industry and the Health and Safety Executive to apprise and carry out such surveys is necessarily limited, the number of such surveys which will be required to be conducted must also necessarily be limited.

The Committee decided that multiplication by a factor of ten of the quantities now set out in the list entitled "Inventories Requiring Notification" (see p. 237, above) would produce a not unmanageable number of sites in the first instance. It was, therefore, decided to use this method of initial assessment even though it gives undue and spurious weight to the inventory at the expense of other factors.

It was for this reason that the Commission urged that the regulations should be prepared in a form which would facilitate the amendment of criteria at appropriate intervals. The proposed regulations require that the responsible person should state ways in which, under faulty conditions, hazardous material might escape containment and they also require that person to estimate the quantity and rate of material release. Further provisions require that the best estimate be made of the effects of an escape, that the probability of an escape occurring be assessed and that the precautions taken to prevent such an occurrence be stated. Hazard surveys must be up-dated as required.

In some cases, a more detailed assessment will be required by the Health and Safety Executive. Such an assessment must include the following information and company procedures:

(i) Management systems and staffing arrangements for the control of hazards.
(ii) Safety systems and procedures for the control of hazards.
(iii) The qualifications, experience and training of the staff concerned.
(iv) The design and operating documentation.
(v) The design and operation of containment and pressure systems.
(vi) The protection of personnel from the effects of loss of containment.
(vii) Emergency plans.
(viii) The reporting of and learning from accidents.

It was recognised by the Committee that these detailed assessments would be costly but requests for such information would be restricted to a minimum and, in most cases, would concern only one or two aspects of the undertaking. The most important hazard survey conducted to date, is the review of potential hazards in the Canvey Island/Thurrock area.

Planning Controls

The Robens Committee considered it desirable to preserve a cordon of safety between potentially hazardous activity and nearby residential areas[24] and saw some relevance in the use of local authority control powers over development, although it saw limits from the point of view of public safety. First, planning controls do not embrace all developments and modifications of use. Second, local authorities often lack the required specialised technical knowledge and resources. Third, the local authority may feel unwilling to impose restrictions because of the importance of the industry to the area in purely economic terms.[25] Because development control procedures could not serve as a main line of defence, it was proposed to augment the powers of local authorities "to take account of the public safety implications of all applications for planning permission and to consult the central Authority responsible for industrial safety in any case where they are in doubt."[26]

The Committee recommended that local authorities should keep their planning control powers in relation to the siting of notifiable installations. Absolute safety, was, it was said, impossible to achieve and some weighing up of the advantages and benefits derived in relation to the risk imposed was desirable. It was important that members of the local community should have an opportunity to make their views known, for which reason the location of a hazardous development should always be a planning matter, whereas regulations concerning containment and control should be enforced by the Health and Safety Executive.

The Committee also recommended that planning control should be executed by local authorities in relation to—

(i) The introduction of potential hazards as part of the development of green field sites.
(ii) The first introduction of potential hazards to existing installations.
(iii) The intensification of hazards at existing installations.

Developers would be obliged to inform planning authorities when installations were notified to the Health and Safety Executive. Technical guidance available to local authorities in the discharge of their planning responsibilities would be improved.[27]

[24] See Report, Cmnd 5034, p. 93, para. 298.
[25] See in this connection: Report of the Tribunal appointed to inquire into the Disaster at Aberfan on October 21, 1966: H.L. 316 H.C. 553 H.M.S.O. (1967).
[26] Report Cmnd. 5034, p. 94, para. 303.
[27] Now the task of the Major Hazards Assessment Unit of the Health and Safety Executive.

The proposed Hazardous Installations (Notification and Survey) Regulations will be placed before Parliament as soon as amendments made necessary by the passage of the EEC Council Directive on Major Industrial Hazards have been agreed.[28] This Council Directive is broadly similar in scope to the proposed regulations, save that its general provision will apply to a lower level of hazard.

[28] Parl. Deb. H.C. (Written Answers), December 4, 1981, col. 249.

CHAPTER 12

SUPRA-NATIONAL SOURCES OF HEALTH AND SAFETY LAW

THE EUROPEAN COMMUNITIES

The European Communities' approach to health and safety law

The regulation of industrial health and safety has played an important part in the work of the European Communities since the foundation of the Coal and Steel Community in 1951, and the foundation of the European Economic Community and the European Atomic Energy Community in 1957. In relation to health and safety in coal mines a separate Mines, Safety and Health Commission was set up to study relevant problems[1] and, on the basis of articles 30–39 of the Euratom treaty, standards of health and protection were established from the coming into force of that treaty.[2] At first, however, the Communities confined themselves to the encouragement of research, the promotion of exchanges of experience and the development of common guidelines in legislation. But the improvement of health and safety was regarded as so important by the signatories to the Treaty that it was included in the preamble of the Euratom and EEC treaties and in Article 3 of EEC treaty whilst, at the same time, it was seen as an essential pre-requisite to the achievement of economic integration. By 1974, however, the Community[3] was annually recording nearly one hundred thousand deaths and twelve million injuries arising from accidents of all types for a workforce of some one hundred and four million. Industrial accidents including occupational diseases, although not the major sector of risk as far as fatal accidents were concerned, represented the largest group of accidents taken as a whole and therefore constituted a priority area for Community concern.

The European Communities' Action Programme on safety and health at work, which was drawn up in order to deal with the problems outlined above, was adopted by the Council of Ministers in a Resolution of June 29, 1978.[4] In its resolution the

[1] O.J. No. C. 487, 1957.
[2] O.J. No. C. 221, 1959.
[3] The three Communities were merged by the Merger or Fusion Treaty on July 1, 1967.
[4] O.J. No. C. 165 1978. It is important to stress that the many actions undertaken within the framework of the Coal and Steel and the Euratom treaties and also aimed at increasing safety and health at work are not covered by this programme.

Council agreed that the following actions could be undertaken up to the end of 1982:

Accident and disease aetiology connected with work—Research

1. Establish, in collaboration with the Statistical Office of the European Communities, a common statistical methodology in order to assess with sufficient accuracy the frequency, gravity and causes of accidents at work, and also the mortality, sickness and absenteeism rates in the case of diseases connected with work.
2. Promote the exchange of knowledge, establish the conditions for close co-operation between research institutes and identify the subjects for research to be worked on jointly.

Protection against dangerous substances

3. Standardize the terminology and concepts relating to exposure limits for toxic substances. Harmonise the exposure limits for a certain number of substances, taking into account the exposure limits already in existence.
4. Develop a preventive and protective action for substances recognized as being carcinogenic, by fixing exposure limits, sampling requirements and measuring methods, and satisfactory conditions of hygiene at the workplace, and by specifying prohibitions where necessary.
5. Establish, for certain specific toxic substances such as asbestos, arsenic, cadmium, lead and chlorinated solvents, exposure limits, limit values for human biological indicators, sampling requirements and measuring methods, and satisfactory conditions of hygiene at the workplace.
6. Establish a common methodology for the assessment of the health risks connected with the physical, chemical and biological agents present at the workplace, in particular by research into criteria of harmfulness and by determining the reference values from which to obtain exposure limits.
7. Establish information notices on the risks relating to and handbooks on the handling of a certain number of dangerous substances such as pesticides, herbicides, carcinogenic substances, asbestos, arsenic, lead, mercury, cadmium and chlorinated solvents.

Prevention of the dangers and harmful effects of machines

8. Establish the limit levels for noise and vibrations at the workplace and determine practical ways and means of

protecting workers and reducing sound levels at places of work. Establish the permissible sound levels of building-site equipment and other machines.

9. Undertake a joint study of the application of the principles of accident prevention and of ergonomics in the design, construction and utilisation of the plant and machinery, and promote this application in certain pilot sectors, including agriculture.
10. Analyse the provisions and measures governing the monitoring of the effectiveness of safety and protection arrangements and organise an exchange of experience in this field.

Monitoring and inspection—improvement of human attitudes

11. Develop a common methodology for monitoring both pollutant concentrations and the measurement of environmental conditions at places of work; carry out intercomparison programmes and establish reference methods for the determination of the most important pollutants. Promote new monitoring and measuring methods for the assessment of individual exposure, in particular through the application of sensitive biological indicators. Special attention will be given to the monitoring of exposure in the case of women, especially of expectant mothers, and adolescents. Undertake a joint study of the principles and methods of application of industrial medicine with a view to promoting better protection of workers' health.
12. Establish the principles and criteria applicable to the special monitoring relating to assistance or rescue teams in the event of accident or disaster, maintenance and repair teams and the isolated worker.
13. Exchange experience concerning the principles and methods of organisation of inspection by public authorities in the fields of safety, hygiene at work and occupational medicine.
14. Draw up outline schemes at a Community level for introducing and providing information on safety and hygiene matters at the workplace to particular categories of workers such as migrant workers, newly recruited workers and workers who have changed jobs.[5]

[5] For details of progress to date on the implementation of this action programme, see COM (80) 370 final: Fourth Progress Report of the Advisory Committee on Safety, Hygiene and Health Protection at Work.

Legal basis for action

The E.E.C. treaty, which is the principal legal basis for Community action, contains two provisions which are *inter alia*, devoted to the promotion of health and safety, *viz.* articles 117 and 118. Measures may also be taken in particular sectors such as agriculture or transport for the removal of barriers to trade on the legal bases appropriate to those sectors, *viz.* articles 43 and 75. These measures may take into account health and safety considerations. In addition, because of the close connection of health and safety with the functioning of the common market the treaty's general legislative powers are used as a basis for action.

The two general bases for action in the treaty are articles 100 and 235. Article 100 refers to the ability of the Council "to issue directives for the approximation of such provisions laid down by law, regulation or administrative action in Member States as directly affect the establishment or functioning of the common market." In so doing the Council must act unanimously on a proposal from the Commission and the Assembly and the Economic and Social Committee must be consulted in the case of directives whose implementation would, in one or more Member States, involve the amendment of legislation. Two points should be noted. First article 100 specifies the instrument to be employed, *viz.* a directive, and, second, the powers of article 100 are available only where the lack of harmonisation of national measures directly affects the establishment or functioning of the common market. For these and possibly other reasons recourse may be had to article 235. Article 235 provides that "if action by the Community should prove necessary to attain, in the course of the operation of the common market, one of the objectives of the Community and the [EEC] Treaty has not provided the necessary powers, the Council shall, acting unanimously on a proposal from the Commission and after consulting the Assembly, take the appropriate measures." The legal instrument used need not, in this case, be a directive and this means may be adopted where the intention is to create new policy.

Legal instruments

The community legislator has a wide choice of instruments which he may use in order to effect this programme. These comprise regulations, directives, decisions, recommendations and opinions.

Regulations: Article 189 of the EEC Treaty provides:

> "A regulation shall have general application. It shall be binding in its entirety and directly applicable in all Member States."

Regulations must be recognised as legal instruments and do not need national implementation—indeed it is impermissible to do so.[6] Regulations are capable of creating individual rights which the Courts must protect.[7] At the time of writing only two regulations have been made in the sphere of Health and Safety at Work; both are of an administrative nature and concern the establishment of a foundation for the improvement of living and working conditions.[8]

Directives and decisions: Article 189 of the EEC Treaty provides:

> "A directive shall be binding, as to the result to be achieved, upon each Member State to which it is addressed, but shall leave to the national authorities the choice of form and methods.
>
> A decision shall be binding in its entirety upon those to whom it is addressed."

The Court of Justice has held that in certain circumstances, at least, directives and decisions might contain directly effective provisions.[9] For this to occur two conditions must be fulfilled. First there must be a clear and precise obligation and secondly it must not require the intervention of any act on the part of the institutions of the Community or the Member States.[10] Individuals may invoke before their national courts, as against the State, such provisions despite the fact that they are contained in directives or decisions rather than regulations. It is also arguable that in some circumstances at least obligations imposed by directives on Member States can create rights in individuals against other individuals (see *Verbond van Nederlandse Ondernemingen* v. *Inspecteur der Invoer rechten en Accinjen*[11] and *Defrenne* v. *Sabena*[12]).

[6] See *Commission* v. *Italy* Case 39/72 [1973] E.C.R. 101.
[7] *Politi* v. *Italian Ministry of Finance* Case 43/71 [1971] E.C.R. 1039.
[8] See Appendix 3, *infra* p. 265.
[9] See *Grad.* v. *Finanzamt Traunstein* Case 9/70 [1970] E.C.R. 825. *Van Duyn* v. *Home Office* Case 41/74 [1974] E.C.R. 1337. and *Rutili* v. *French Minister of the Interior* Case 36/75 E.C.R. 1219.
[10] See E.E.C. Directives and the Law G.L. Close Vol. 2. Planning Law for Industry International Bar Association Cambridge 1981.
[11] Case 57/76 [1977] E.C.R. 113.
[12] Case 43/75 [1976] E.C.R. 455. See Josephine Steiner, "Direct Applicability—A Chameleon Concept," Vol. 98 L.Q.R. (April 1982), pp. 229–248, for an analysis and review of direct application and direct consequences.

Directives have been made in a large number of diverse areas in the health and safety field some of the more important of which are listed below:
Safety signs; safety requirements for tower cranes for building work; on the protection of workers from harmful exposure to chemical, physical and biological agents at work; on the approximation of laws relating to powered industrial trucks; on major accident hazards of certain industrial activities; on the protection of workers from harmful exposure to metallic lead and its ionic compounds at work; on the protection of the public against the danger of microwave radiation; on the health of workers exposed to vinyl chloride monomer; on inspection procedures for pressure vessels; on the classification, packaging and labelling of dangerous substances; on the use of electrical equipment in potentially explosive atmospheres; on the use of tractors and agricultural machinery; and on wire ropes and toxic products.[13]

Other measures: Recommendations and opinions have no binding force but may be indicative of important policy orientations and Member States may take such measures into account when enacting national legislation or when making administrative directions. Similarly resolutions and declarations by the Council do not produce legal results but contain important statements of policy which serve as a guide to Community Institutions and Member States.

This legislative activity by the European Communities was not without its impact on the rôle of the Health and Safety Commission in its task of proposing health and safety regulations, approved codes of practice and guidance notes in the United Kingdom. When the directive on safety signs and safety information was proposed by the Commission in 1976,[14] the Health and Safety Executive argued[15] that if such a directive were made there would be insufficient time for the consultation procedure to take place before regulations had to be made and, also, that their programme would be harmed if this proposal had to be dealt with quickly and out of turn.

Whilst on the one hand in relation to the Directive on Major Accident Hazards of certain industrial activities the Health and Safety Executive argued,[16] *inter alia*, that the proposals went wider and deeper than was necessary and that the notification

[13] See further Appendix 3, *infra* p. 265.
[14] O.J. No. C. 96 1976.
[15] See Select Committee of the House of Lords on the European Communities 52nd Report Session 1975–1976.
[16] See Select Committee of the House of Lords on the European Communities 33rd Report Session 1979–1980.

procedures might delay the implementation of new processes unnecessarily. On the other hand the memorandum submitted by the Chairman of the Health and Safety Commission on the two asbestos directives[17] showed that in many cases the HSC would prefer standards higher than those which it was suggested should be imposed. Furthermore, the HSC was concerned in case difficulties might be created if the United Kingdom decided on this course of action since the draft directive on the marketing and use of asbestos was being put forward under Article 100 of the EEC Treaty[18] and the U.K.'s action unilaterally raising standards might be seen as placing obstacles in the path of fair competition.

In Health and Safety Commission Report 1976–1977 the complaint was voiced that European Community directives offered no choice in their implementation.[19] During that year a Pressure Vessels directive had to be implemented by making regulations in advance of the Commission's own intentions. They had intended to make regulations with wider implications, and the general point was made that improved arrangements for forecasting progress with directives affecting the Commission's responsibilities would be welcomed. However, difficulties expressed in the Report of the Health and Safety Commission for 1978–79[20] concerning the compatibility of the two formulating bodies, *viz.* the Commission of the European Communities and the Health and Safety Commission appear now to have been resolved.

THE INTERNATIONAL LABOUR ORGANISATION

Tripartite Structure

Formed in 1919 as an autonomous body associated with the League of Nations (and with Great Britain as a founder member), the ILO survived the demise of the ill-fated League, and later hitched its star to the United Nations, of which it has been a specialised agency since 1946. Based from the beginning on the inseparability of peace and social justice, the aims and principles of the ILO were re-affirmed and strengthened by the Declaration of Philadelphia (1944), which brought the Organisation into the struggle against poverty and insecurity. Since

[17] To the Select Committee of the House of Lords on the European Communities 17th Report Session 1980–1981.
[18] There was a danger said the Chairman of the Health and Safety Commission that this could be construed by the European Court as suspending Article 36 Treaty of Rome.
[19] Health and Safety Commission Report 1976–1977 HMSO 1978 at page 11.
[20] Health and Safety Commission Report 1978–1979 HMSO 1980 at page 4.

its inception in 1919, the ILO has undergone some fundamental changes. Whilst it has carried on with its standard-setting role as a "World Industrial Parliament," it has, in recent decades, involved itself in technical programmes on a large scale, working with other UN bodies such as the United Nations Development Programme and the World Health Organisation. Formed in Europe from the richer industrialised nations it has turned its attention to the developing countries where the need of social amelioration is greatest.

For our purposes, the ILO is of interest for the international labour standards which it promulgates, amongst which are those related to Article III(g) of the Philadelphia Declaration—"adequate protection for the life and health of workers in all occupations." ILO standards take two main forms—Conventions and Recommendations.

A two-thirds majority of those voting at the Conference of the ILO is required for the adoption of a Convention, which the national members must then bring before the competent authority for the enactment of legislation or other action.[21] When ratified by a Member State the latter is under a binding international obligation to ensure that its law and practice is in conformity with the Convention. Recommendations on the other hand are not "binding" in this sense, but provide guides for national action. Thus our law on unfair dismissal, which was introduced in 1971, owes a good deal to Recommendation 119 of the ILO, showing that Member States seek to implement Recommendations even though these cannot be ratified in the same way as a Convention. In addition to these two types of standards, considerable influence can be exercised by quasi-norms such as resolutions, reports, requests and memoranda.

Two features of the ILO are its tripartite structure and method of operation, and the machinery for achieving conformity with ILO standards. From the outset, the ILO has embraced "tripartism" in its structure. The Conference, which promulgates the international labour legislation, is itself composed of representatives of government, employers and workers each of whom is represented in the permanent secretariat at Geneva. In all of its activities the Organisation seeks to proceed on the basis of consensus between these three constituencies so that, for example, a technical mission to a developing country will work with government, employers and workers' representatives.

Secondly, whilst the ILO lacks the close integration of a regional organisation such as the EEC, or the right of individual petition available through the European Convention on Human

[21] Normally a Convention requires at least two ratifications to become operative.

Rights, it would be a mistake to assume that its promulgated standards are merely precatory expressions of hope. The system of reporting from Member States, the investigations of committees, particularly the Committee of Experts, and the procedures for dealing with complaints and representations, show that the system is intended to have normative effect. Under Article 24 of the ILO Constitution, complaints relating to the non-observance of a Convention may be lodged by industrial associations of employers or workers. Although governmental implementation of Conventions leaves much to be desired, it is undeniable that the Conventions do have considerable regulative effect, even in countries with a scrupulous regard for international obligations, such as the U.K. Because of the diverse national conditions within which the ILO norms have to operate, the system is designed to allow considerable flexibility in national implementation, *e.g.* it accepts lower minimum age standards for certain countries, or the exclusion of commercial premises from labour inspection.

Labour standards

As will be seen from the Conventions and Recommendations listed on page 251, the standards have dealt with particular occupational safety and health problems, such as particular workers, industries or risks.

A new comprehensive approach

In contrast to the hitherto piecemeal approach, the 1981 Conference had before it a proposed Convention concerning Occupational Safety and Health and the Working Environment. Introducing this new general standard, the Chairman of the Conference justified the need for a "comprehensive document that would determine the fundamental forms and direction of activities aimed at the protection of man at work" by reference to the dynamic progress of technology which affects the working environment by creating new hazards and intensifying old hazards—the "hundreds of new chemical compounds introduced into the market every year, electro-magnetic and laser radiations, infra- and ultra-sounds." It was her conviction that "all countries were developing countries" so far as health and safety at work and the working environment were concerned. Just as Britain has put the coping stone of a general Act to her occupational safety and health edifice, so the ILO has moved towards a comprehensive standard which will apply to all branches of economic activity and to all workers (including

LIST OF INSTRUMENTS CONCERNING OCCUPATIONAL SAFETY AND HEALTH AND THE WORKING ENVIRONMENT ADOPTED BY THE INTERNATIONAL LABOUR CONFERENCE SINCE 1919

Year	Convention	Recommendation
1921	13. White Lead (Painting)	
1929	27. Marking of Weight (Packages Transported by Vessels)	
1937	62. Safety Provisions (Building)	53. Safety Provisions (Building)
1946	73. Medical Examination (Seafarers)	79. Medical Examination of Young Persons
	77. Medical Examination of Young Persons (Industry)	
	78. Medical Examination of Young Persons (Non-industrial Occupations)	
1947	81. Labour Inspection	81. Labour Inspection
		82. Labour Inspection (Mining and Transport)
1949	92. Accommodation of Crews (Revised)	
1953		97. Protection of Workers' Health
1958		105. Ships' Medicine Chests
		106. Medical Advice at Sea
1959	113. Medical Examination (Fishermen)	112. Occupational Health Services
1960	115. Radiation Protection	114. Radiation Protection
1963	119. Guarding of Machinery	118. Guarding of Machinery
1964	120. Hygiene (Commerce and Offices)	120. Hygiene (Commerce and Offices)
	121. Employment Injury Benefits	121. Employment Injury Benefits
1965	124. Medical Examination of Young Persons (Underground Work)	
1967	127. Maximum Weight	128. Maximum Weight
1969	129. Labour Inspection (Agriculture)	133. Labour Inspection (Agriculture)
1970	133. Accommodation of Crews (Supplementary Provisions)	140. Crew Accommodation (Air Conditioning)
		141. Crew Accommodation (Noise Control)
	134. Prevention of Accidents (Seafarers)	142. Prevention of Accidents (Seafarers)
1971	136. Benzene	144. Benzene
1974	139. Occupational Cancer	147. Occupational Cancer
1977	148. Working Environment (Air Pollution, Noise and Vibration)	156. Working Environment (Air Pollution, Noise and Vibration)
1979	152. Occupational Safety and Health (Dock Work)	160. Occupational Safety and Health (Dock Work)

public employees) in the branches of economic activity covered. Each Member State in consultation with representative organisations of employers and workers will need to formulate, implement and periodically review a coherent national policy on occupational safety, occupational health (including well-being) and the working environment. The inclusion of "well-being" shows that a wide view is taken of health and safety. The aim of the policy is to prevent accidents and injury to health (once again including well-being) arising out of, linked with or occurring in the course of work by minimising, *so far as is reasonably practicable*, the causes of hazards inherent in the working environment (*emphasis added*). It is interesting to see the British formula of reasonable practicability appearing in this proposed new standard. The policy referred to above must take account of the following matters (set out in Article 5):

(*a*) design, testing, integration, choice, substitution, installation, arrangement, use and maintenance of the material elements of work (working environment, machinery and equipment, substances and chemical, physical and biological agents used, work processes);

(*b*) relationships between the material elements of work and the persons who carry out or supervise the work, and adaptation of machinery, equipment, working time, organisation of work and work processes to the physical and mental capacities of the workers;

(*c*) training, retraining, qualifications and motivation of persons involved, in one capacity or another, in the achievement of adequate levels of safety and heath;

(*d*) communication and co-operation, at the levels of the working group and the undertaking, as well as at the national level.

The policy must indicate the respective functions and responsibilities of public authorities, employers, workers and others and take account of the complementary character of such responsibilities and of national practice and conditions (Article 6). Enforcement will be secured "by an adequate and appropriate system of inspection" backed up by "adequate penalties for violations of the laws and regulations" (Article 9). Measures must be taken to provide guidance to employers and workers to enable them to comply with their legal obligations (Article 10).

Under Article 11 of the proposed Convention the measures to give effect to the policy referred to above shall progressively include the assumption of the following functions by the competent authority—

(*a*) the determination, where the nature and degree of hazards so require, of conditions governing the construction and lay-out of undertakings, the commencement of their operations, major alterations affecting them and changes in their purposes, as well as the application of procedures defined by the competent authorities;

(*b*) the determination of substances and agents exposure to which is to be prohibited, limited or made subject to authorisation or control by the competent authority or authorities; health hazards due to the simultaneous use of several substances and simultaneous effects shall be taken into consideration;

(*c*) the establishment and application of procedures for the notification of occupational accidents and diseases, by employers and, when appropriate, insurance institutions and others directly concerned, and the production of annual statistics on occupational accidents and diseases;

(*d*) the holding of inquiries, where cases of occupational accidents, occupational diseases and all other injuries to health which arise in the course of or in connection with work appear to reflect situations which are particularly serious;

(*e*) the publication, annually, of information on measures taken in pursuance of the policy referred to in Article 4 and on occupational accidents, occupational diseases and all other injuries to health which arise in the course of or in connection with work.

Article 12 is evocative of our section 6 (see p. 92, above) in requiring that measures shall be taken to ensure that those who design, manufacture, import, provide or transfer machinery, equipment, substances or agents for occupational use shall—

(*a*) satisfy themselves that, in so far as is reasonably practicable, the machinery, equipment, substance or agent does not entail dangers for the safety and health of those using it correctly;

(*b*) make available information concerning the correct installation and use of machinery and equipment and the correct use of substances and agencies;

(*c*) undertake studies and research or otherwise keep abreast of the scientific and technical knowledge necessary to comply with (*a*) and (*b*) above.

Article 13 requires the taking of measures to promote in a manner appropriate to national practice and conditions the inclusion of occupational safety and health and the working environment at all levels of education and training, including those of higher technical, medical and professional institutes, and that in a manner meeting the training needs of all workers.

The new Convention imposes obligations at the level of the undertaking. Article 15 requires employers to ensure so far as is reasonably practicable that workplaces, machinery, equipment and processes under their control are safe and without risks to health, whilst Article 16 imposes further requirements as to emergencies and accidents (including adequate first-aid arrangements). Those familiar with the British law and practice since 1974 will be able to spot those of our provisions which are equivalent or similar to those in Article 17 which requires arrangements at the level of the undertaking under which—

(*a*) workers, in the course of performing their work, co-operate in the fulfilment by their employer of the obligations placed upon him;

(*b*) representatives of workers in the undertaking co-operate with the employer in the field of occupational safety and health;

(*c*) representatives of workers in an undertaking are given adequate information on measures taken by the employer to secure their occupational safety and health (including well-being) and may consult their representative organisations about such information provided they do not disclose commercial secrets;

(*d*) workers and their representatives in the undertaking are given appropriate training in occupational safety and health;

(*e*) workers and their representative organisations in an undertaking are enabled to inquire into, and are consulted by the employer on, all aspects of occupational safety and health (including well-being) associated with their work; for this purpose technical advisers may, by mutual agreement, be brought in from outside the undertaking.

Article 18 emphasises that co-operation between management and workers shall be an essential element of organisational and other measures taken in pursuance of Articles 15 to 17 (which, in the words of Lon L. Fuller, might be said to embody the "morality of aspiration" rather than the present position in some enterprises). Article 19 confers a new and dramatic right upon workers (which may have been drawn from American law)[22] when it states that—

(a) A worker shall have the right to cease work when he judges the work to involve imminent and serious danger to his life or health, on condition that he reports the cessation of work immediately to his employer, or to a workers' safety delegate.

(b) A worker shall not be dismissed or otherwise prejudiced by reason of having ceased to work in accordance with (a) above nor shall he be held responsible for any damage or liability arising from the cessation of work, as measured

[22] Although the Supreme Court in *Marshall* v. *Daniell Construction Co.* (see *Monthly Labor Review*, March 1979, p. 61) denied review of a court decision invalidating a "right to refuse" regulation of the Labor Department, such a regulation, where it entitled the employee to refuse work exposing him to a real risk of serious injury or even death, was later upheld (see *Monthly Labor Review*, April 1980, p. 57).

from the time the work ceases until a decision is made to resume work.

Our law does not statutorily guarantee a right to cease work for this reason, although it is a common law implied term that the employer cannot subject the worker to a risk to health or limb (not expressly undertaken by the worker) outside the scope of his contractual employment; but this provision would merely give the worker a limited right to damages for wrongful dismissal at common law if he were to be dismissed for refusing to run such a risk. Our law of unfair dismissal (but not the law on "action short of dismissal") is capable of affording some protection under (b) above, although the words "shall not be dismissed" would seem to indicate that dismissal followed by compensation would not suffice.

The Sixty-Seventh General Conference of 1981 also proposed a Recommendation on the same subject-matter.

APPENDIX 1

Regulations made under provisions of Part 1 of the Health and Safety at Work etc. Act 1974, and statutory instruments relevant to Part 1 made under the Provisions of Part IV of the Act.

A. General

The Health and Safety at Work etc. Act 1974 (Commencement No. 1) Order 1974 No. 1439.

The Anthrax Prevention Act 1919 (Repeals and Modifications Regulations 1974 No. 1775.

The Factories Act 1961 (Enforcement of Section 135) Regulations 1974 No. 1776

The Docks and Harbours Act 1966 (Modification) Regulations 1974 No. 1820

The Radioactive Substance Act 1948 (Modification) Regulations 1974 No. 1821

The Hydrogen Cyanide (Fumigation Act 1937 (Repeals and Modifications) Regulations 1974 No. 184

The Celluloid and Cinematograph Film Act 1922 (Repeals and Modifications) Regulations 1974 No. 184

The Explosives Acts 1875 and 1923 etc. (Repeals and Modifications) Regulations 1974 No. 1885

The Boiler Explosions Acts 1882 and 1890 (Repeals and Modifications) Regulations 1974 No. 1886

The Truck Acts 1831 to 1896 (Enforcement) Regulations 1974 No. 1887

The Industrial Tribunals (Improvement and Prohibition Notices Appeals) Regulations 1974 No. 192.

The Industrial Tribunals (Improvement and Prohibition Notices Appeals) (Scotland) Regulations 1974 No. 1926

The Factories Act 1961 etc. (Repeals and Modifications) Regulations 1974 No. 1941

The Petroleum (Regulation) Acts 1928 and 1936 (Repeals and Modifications) Regulations 1974 No. 1942

The Offices, Shops and Railway Premises Act 1963 (Repeals and Modifications) Regulations 1974 No. 1943

The Pipe-lines Act 1962 (Repeals and Modifications) Regulations 1974 No. 1986

The Coal Industry Nationalisation Act 1946 (Repeals) Regulations 1974 No. 2011

The Ministry of Fuel and Power Act 1945 (Repeal) Regulations 1974 No. 2012

The Mines and Quarries Acts 1954 to 1971 (Repeals and Modifications) Regulations 1974 No. 2013

The Health and Safety Licensing Appeals (Hearings Procedure) Rules 1974 No. 2040

The Nuclear Installations Act 1965 etc. (Repeals and Modifications) Regulations 1974 No. 2056

The Health and Safety Licensing Appeals (Hearings Procedure) (Scotland) Rules 1974 No. 2068

The Explosives Acts 1875 and 1923 etc. (Repeals and Modifications) (Amendment) Regulations 1974 No. 2166 Clean Air Enactments (Repeals and Modifications) Regulations 1974 No. 2170

The Agriculture (Poisonous Substances) Act 1952 (Repeals and Modifications) Regulations 1975 No. 45

The Agriculture (Safety, Health and Welfare Provisions) Act 1956 (Repeals and Modifications) Regulations 1975 No. 46

The Health and Safety (Agriculture) (Poisonous Substances) Regulations 1975 No. 282

The Protection of Eyes (Amendment) Regulations 1975 No. 303

The Health and Safety Inquiries (Procedure) Regulations 1975 No. 335

The Offices, Shops and Railway Premises Act 1963 (Repeals) Regulations 1975 No. 1011

The Factories Act 1961 (Repeals) Regulations 1975 No. 1012

The Mines and Quarries Acts 1954 to 1971 (Repeals and Modifications) Regulations 1975 No. 1102

The Coal Mines (Respirable Dust) Regulations 1975 No. 1433

The Employers' Health and Safety Policy Statements (Exception) Regulations 1975 No. 1584

The Conveyance of Explosives by Road (Special Case) Regulations 1975 No. 1621

The Baking and Sausage Making (Christmas and New Year) Regulations 1975 No. 1695 (spent)

The Coal Mines (Precautions against Inflammable Dust) Temporary Provisions Regulations 1976 No. 881 (spent)

The Operations at Unfenced Machinery (Amendment) Regulations 1976 No. 955

The Health and Safety Inquiries (Procedure) (Amendment) Regulations 1976 No. 1246

The Health and Safety (Agriculture) (Miscellaneous Repeals and Modifications) Regulations 1976 No. 1247

The Baking and Sausage Making (Christmas and New Year) Regulations 1976 No. 1908 (spent)

The Fire Certificates (Special Premises) Regulations 1976 No. 2003

The Factories Act 1961 etc. (Repeals) Regulations 1976 No. 2004

The Offices, Shops and Railway Premises Act 1963 etc. (Repeals) Regulations 1976 No. 2005

The Fire Precautions Act 1971 (Modifications) Regulations 1976 No. 2007

The Mines and Quarries (Metrication) Regulations 1976 No. 2063

The Safety Representatives and Safety Committees Regulations 1977 No. 500

The Health and Safety (Enforcing Authority) Regulations 1977 No. 746

The Coal Mines (Precautions against Inflammable Dust) Amendment Regulations 1977 No. 913

The Explosives (Registration of Premises) Variation of Fees Regulations 1977 No. 918

The Coal and Other Mines (Electricity) (Third Amendment) Regulations 1977 No. 1205

The Health and Safety at Work etc. Act 1974 (Application outside Great Britain) Order 1977 No. 1232

The Acetylene (Exemption) Order 1977 No. 1798 (spent)

The Baking and Sausage Making (Christmas and New Year) Regulations 1977 No. 1841 (spent)

The Packaging and Labelling of Dangerous Substances Regulations 1978 No. 209

The Explosives (Licensing of Stores) Variation of Fees Regulations 1978 No. 270

The Petroleum (Regulation) Acts 1928 and 1936 (Variation of Fees) Regulations 1978 No. 635

The Health and Safety (Genetic Manipulation) Regulations 1978 No. 752

The Coal Mines (Respirable Dust) (Amendment) Regulations 1978 No. 807

The Factories (Standards of Lighting) Revocation Regulations 1978 No. 1126

The Baking and Sausage Making (Christmas and New Year) Regulations 1978 No. 1516 (spent)

The Coal and other Mines (Metrication) Regulations 1978 No. 1648

The Hazardous Substances (Labelling of Road Tankers) Regulations 1978 No. 1702

The Compressed Acetylene (Importation) Regulations 1978 No. 1723

The Mines and Quarries Act 1954 (Modification) Regulations 1978 No. 1951

The Mines (Precautions Against Inrushes) Regulations 1979 No. 318 (coming into operation 9 April 1979)

The Petroleum (Consolidation) Act 1928 (Enforcement) Regulations 1979 No. 427

The Coal and Other Mines (Electric Lighting for Filming) Regulations 1979 No. 1203

The Baking and Sausage Making (Christmas and New Year) Regulations 1979 No. 1298 (spent)

The Explosives Act 1875 (Exemptions) Regulations 1979 No. 1378

The Health and Safety (Fees for Medical Examinations) Regulations 1979 No. 1553

The Notification of Accidents and Dangerous Occurrences Regulations 1980 No. 804

The Health and Safety (Leasing Arrangements) Regulations 1980 No. 907

The Coal and Other Mines (Fire and Rescue) (Amendment) Regulations 1980 No. 942

The Agriculture (Tractor Cabs) (Amendment) Regulations 1980 No. 1036

The Petroleum (Consolidation) Act 1928 (Conveyance by Road Regulations Exemptions) Regulations 1980 No. 1100

The Mines and Quarries (Fees for Approvals) Regulations 1980 No. 1233

The Control of Lead at Work Regulations 1980 No. 1248

The Celluloid and Cinematograph Film Act 1922 (Exemptions) Regulations 1980 No. 1314

The Safety Signs Regulations 1980 No. 1471

The Baking and Sausage Making (Christmas and New Year) Regulations 1980 No. 1576 (spent)

The Health and Safety (Animal Products) (Metrication) Regulations 1980 No. 1690

The Health and Safety (Enforcing Authority) (Amendment) Regulations 1980 No. 1744

The Chemical Works (Metrication) Regulations 1981 No. 16

The Mines and Quarries (Fees for Approvals) (Amendment) Regulations 1981 No. 270

The Health and Safety (Fees for Medical Examinations) Regulations 1981 No. 334

The Diving Operations at Work Regulations 1981 No. 399

The Baking and Sausage Making (Christmas and New Year) Regulations 1982 No. 1498

The Petroleum-Spirit (Plastic Containers) Regulations 1982 No. 630

The Hydrogen Cyanide (Fumigation of Buildings) (Amendment) Regulations 1982 No. 695

The Offices, Shops and Railway Premises Act 1963 etc. (Metrication) Regulations 1982 No. 827

The Pottery (Health etc.) (Metrication) Regulations 1982 No. 877.

B. Regulations made relating to particular mines:

The Shilbottle Mine (Endless Rope Haulage) (Revocation) Regulations 1975 No. 1078

The Polmaise Mine (Nos 3 and 5 Shafts) Regulations 1975 No. 1079

The Gartmorn Mine (Precautions against Inrushes) (Amendment) Regulations 1975 No. 1394

The Westoe Mine (St. Hilda Shaft) Regulations 1975 No. 1395

The Haig Mine (Thwaites Shaft) Regulations 1975 No. 1519

The Killoch Mine (Electric Lighting) Regulations 1975 No. 1632

The Bagworth Mine (Precautions against Inrushes) (Revocation) Regulations 1975 No. 1633

The Bolsover Mine (Electric Lighting) Regulations 1975 No. 1676

The Ellistown Mine (Electric Lighting) Regulations 1975 No. 1677

The Parsonage Mine (Electric Lighting) Regulations 1975 No. 1678

The Brodsworth Mine (Electric Lighting) Regulations 1975 No. 1679

The Daw Mill Mine (Electric Lighting) Regulations 1975 No. 1819

The Rawdon Mine (Electric Lighting) Regulations 1975 No. 1820

The Hapton Valley Mine (Electric Lighting) Regulations 1975 No. 1821

The Newdigate Mine (Electric Lighting) Regulations 1975 No. 1822

The Chatterley-Whitfield Mine (Electric Lighting) Regulations 1975 No. 1823

The Harworth Mine (Nos. 1 and 2 Shafts) Regulations 1975 No. 1886
The Dinnington Main Mine (No. 1 Shaft) Regulations 1975 No. 2053
The Ireland Mine (Endless Rope Haulage) Regulations 1975 No. 2218
The Baddesley Mine (Endless Rope Haulage) Regulations 1976 No. 27
The Birch Coppice Mine (Endless Rope Haulage) Regulations 1976 No. 28
The High Moor Mine (Endless Rope Haulage) Regulations 1976 No. 43
The Warsop Main Mine (Endless Rope Haulage) Regulations 1976 No. 44
The Whitwick Mine (Electric Lighting) Regulations 1976 No. 80
The Markham Main Mine (Electric Lighting) Regulations 1976 No. 81
The Snibston Mine (Endless Rope Haulage) Regulations 1976 No. 82
The Desford Mine (Electric Lighting) Regulations 1976 No. 130
The Gedling Mine (No. 1 Downcast Shaft) Regulations 1976 No. 156
The Hem Heath Mine (Electric Lighting) Regulations 1976 No. 254
The Birch Coppice Mine (Electric Lighting) Regulations 1976 No. 255
The Monktonhall Mine (Endless Rope Haulage) Regulations 1976 No. 479
The Snibston Mine (Diesel Vehicles) Regulations 1976 No. 480
The Markham Mine (Endless Rope Haulage) Regulations 1976 No. 481
The Bogside Mine (Endless Rope Haulage) Regulations 1976 No. 482
The Polkemmet Mine (Endless Rope Haulage) Regulations 1976 No. 483
The Arkwright Mine (Endless Rope Haulage) Regulations 1976 No. 484
The Bolsover Mine (Endless Rope Haulage) Regulations 1976 No. 485
The Shirebrook Mine (Endless Rope Haulage) Regulations 1976 No. 486
The Westthorpe Mine (Endless Rope Haulage) Regulations 1976 No. 487
The Daw Mill Mine (Endless Rope Haulage) Regulations 1976 No. 556
The South Leicester Mine (Electric Lighting) Regulations 1976 No. 696
The Whitwell Mine (Endless Rope Haulage) Regulations 1976 No. 967
The Blidworth Mine (Electric Lighting) Regulations 1976 No. 999
The Thurcroft Main Mine (No. 1 Shaft) Regulations 1976 No. 1014
The Wolstanton Mine (Electric Lighting) Regulations 1976 No. 1444
The Seafield Mine (Endless Rope Haulage) Regulations 1976 No. 1445
The Solsgirth Nos. 1 & 2 Mine (Endless Rope Haulage) Regulations 1976 No. 1706
The Killoch Mine (Endless Rope Haulage) Regulations 1976 No. 1608
The Bilstone Glen Mine (Endless Rope Haulage) Regulations 1976 No. 1609
The New Hucknall Mine (Endless Rope Haulage) Regulations 1976 No. 1610
The Brynlliw Mine (No. 2 Upcast Shaft) Regulations 1976 No. 1611
The Cardowan Mine (Endless Rope Haulage) Regulations 1976 No. 1612
The Yorkshire Main Mine (Endless Rope Haulage) Regulations 1976 No. 1654
The Frances Mine (Endless Rope Haulage) Regulations 1976 No. 1655
The Daw Mill Mine (No. 1 Downcast Shaft) Regulations 1976 No. 1732
The High Moor Mine (Diesel Vehicles) Regulations 1976 No. 1733
The Markham Mine (Diesel Vehicles) Regulations 1976 No. 1734
The Bolsover Mine (No. 2 Shaft) Regulations 1976 No. 1735

The Blackdene Mine (Storage Battery Locomotives) (Amendment) Regulations 1976 No. 1827

The Comrie Mine (No. 1 Shaft) (Automatic Shaft Signalling) Regulations 1976 No. 2045

The Bentinck Mine (Diesel Engined Stone Dusting Machine) Regulations 1976 No. 2046

The Birch Coppice Mine (No. 3 (Wood End) Shaft) Regulations 1976 No. 2047

The Linby Mine (Electric Lighting) Regulations 1976 No. 2048

The Welbeck Mine (Winding) Regulations 1976 No. 2049

The Blidworth Mine (Winding) Regulations 1976 No. 2050

The Brodsworth Mine (Endless Rope Haulage) Regulations 1976 No. 2051

The High Moor Mine (Cable Reel Shuttle Cars) Regulations 1976 No. 2052

The Thoresby Mine (Electric Lighting for Cinematography) Regulations 1976 No. 2056

The Winsford Rock Salt Mine (No. 3 Shaft) Regulations 1976 No. 2075

The Markham Mine (Shafts) Regulations 1976 No. 2087

The Teversal Mine (Electric Lighting for Cinematography) Regulations 1977 No. 38

The Cadeby Mine (Electric Lighting Regulations 1977 No. 201

The Markham Mine (Electric Lighting) Regulations 1977 No. 202

The Bentley Mine (Endless Rope Haulage) Regulations 1977 No. 203

The Murton Mine (Endless Rope Haulage) Regulations 1977 No. 225

The Yorkshire Main Mine (Well Shaft) Regulations 1977 No. 226

The Rossington Mine (Endless Rope Haulage) Regulations 1977 No. 483

The Blaenavon Mine (Electric Lighting for Cinematography) Regulations 1977 No. 484

The Goldthorpe/Highgate Mine (Endless Rope Haulage) Regulations 1977 No. 658

The Ireland Mine (Diesel Vehicles) Regulations 1977 No. 735

The Baddesley Mine (No.'s 1 and 2 Upcast Shafts) Regulations 1977 No. 736

The Dinnington Main Mine (Revocation of Special Regulations) Regulations 1977 No. 737

The Ollerton Mine (No. 1 Shaft) Regulations 1977 No. 738

The Bentley Mine (Electric Lighting for Cinematography) Regulations 1977 No. 879

The Hickleton Mine (Electric Lighting for Cinematography) Regulations 1977 No. 880

The Blaenserchan Mine (Diesel Vehicles) Regulations 1977 No. 917

The Allerton Bywater Mine (Electric Lighting for Cinematography) Regulations 1977 No. 967

The Rothwell Mine (Electric Lighting for Cinematography) Regulations 1977 No. 968

The Silverhill Mine (No. 1 Downcast Shaft) Regulations 1977 No. 1045

The Panallta Mine (Electric Lighting for Cinematography) Regulations 1977 No. 1195

The Daw Mill Mine (No. 2 Upcast Shaft) Regulations 1977 No. 1196

The Mount Wellington Mine (Winding) Regulations 1977 No. 1384
The Hatfield/Thorne Mine (No. 2 Shaft) Regulations 1977 No. 1487
The Welbeck Mine (Nos. 1 and 2 Shafts) Regulations 1977 No. 1549
The Parsonage Mine (Endless Rope Haulage) Regulations No. 1593
The Silverhill Mine (Nos. 1 and 2 Shafts) Regulations 1977 No. 1636
The East Hetton Mine (Hutton Shaft) Regulations 1977 No. 1661
The Jubilee Drift Mine (Electric Lighting for Cinematography) Regulations 1977 No. 1662
The Hapton Valley Mine (Diesel Vehicles) Regulations 1977 No. 1696
The Treeton Mine (Refuge Holes) Regulations 1977 No. 1758
The Golborne Mine (Endless Rope Haulage) Regulations 1977 No. 1855
The Bolsover Mine (Cable Reel Shuttle Cars) Regulations 1977 No. 2035
The Cynheider/Pentremawr Mine (Escape Breathing Apparatus) (Amendment) Regulations 1977 No. 2036
The Cwmgwili Mine (Escape Breathing Apparatus) Regulations 1977 No. 2106
The Goldthorpe/Highgate Mine (Precautions Against Inrushes) Regulations 1977 No. 2171
The Gascoigne Wood Mine (Refuge Holes) Regulations 1978 No. 33
The Thoresby Mine (Cable Reel Load — Haul — Dump Vehicles) Regulations 1978 No. 119
The Sherburn No. 2 Mine (Precautions Against Inrushes) Regulations 1978 No. 411
The Sallet Hole Nos. 1 and Mines (Diesel Vehicles) Regulations 1978 No. 761
The Trelewis Drift Mine (Diesel Vehicles) Regulations 1978 No. 1376
The Bentinck Mine (Electric Lighting) Regulations 1978 No. 1476
The West Cannock No. 5 Mine (Electric Lighting for Cinematography) Regulations 1978 No. 1477
The Ackton Hall Mine (Cable Reel Load—Haul—Dump Vehicles) Regulations 1978 No. 1539
The Daw Mill Mine (Refuge Holes) Regulations 1978 No. 1815
The Ellington Mine (No. 3 Shaft) Regulations 1979 No. 67
The Hucknall Mine (No. 5 Shaft) Regulations 1979 No. 288
The Wistow Mine (Electric Lighting for Cinematography) Regulations 1979 No. 436 (revoked by SI 1979/1203)
The Easington Mine (Northshaft) Regulations 1979 No. 585
The Oakdale Mine (Electric Lighting) Regulations 1979 No. 701
The Rufford Mine (No. 1 Shaft) Regulations 1979 No. 983
The Ellington Mine (Electric Lighting) Regulations 1979 No. 984
The Easington Mine (Electric Lighting) Regulations 1979 No.1292
The High Moor Mine (Cable Reel Load — Haul — Dump Vehicles) Regulations 1979 No. 1293
The Thoresby Mine (No. 2 Shaft) Regulations 1979 No. 1413
The Yorkshire Main Mine (Friction Winding) Regulations 1979 No. 1491
The Boulby Mine (Diesel Vehicles) Regulations 1979 No. 1532
The Scraithole Mine (Storage Battery Locomotives) Regulations 1979 No. 1658

The Whitwell Mine (Teleplatform Haulage and Refuse Holes) Regulations 1979 No. 1769

The Murton Mine (Friction Winding) Regulations 1980 No. 68

The Ollerton Mine (Nos. 1 and 2 Shafts) Regulations 1980 No. 260

The Markham Mine (No. 1 Shaft) Regulations 1980 No. 261

The Bilsthorpe Mine (No. 1 Shaft) Regulations 1980 No. 262

The Sallet Hole No. 2 Mine (Storage Battery Locomotives) Regulations 1980 No. 1203

The Lynemouth Mine (Electric Lighting) (Regulations) 1980 No. 1395

The Manton Mine (Electric Lighting) Regulations 1980 No. 1396

The Vane Tempest Mine (Electric Lighting) Regulations 1980 No. 1397

The Yew Tree Mine (Storage Battery Locomotives) Regulations 1980 No. 1405

The Harworth Mine (Cable Reel Load — Haul — Dump Vehicles) Regulations 1980 No. 1474

The Point of Ayr Mine (Diesel Vehicles) Regulations 1980 No. 1705

The Sherburn No. 2 Mine (Diesel Vehicles) Regulations 1980 No. 1891

APPENDIX 2

Codes of Practice Approved by the Health and Safety Commission

Safety Representatives and Safety Committees

Time off for the Training of Safety Representatives

Control of lead at work

Asbestos: Insulation and Coating

Health and Safety (First Aid) Regulations 1981

Classification of Dangerous Substances for Conveyance in Road Tankers and Tank Containers

APPENDIX 3

European Communities' Legislation and United Kingdom Implementation

	EUROPEAN COMMUNITIES' LEGISLATION				*UNITED KINGDOM LEGISLATION*		
	Official Journal Reference	Principal Regulation/ Directive	Modification	Subject Matter	England & Wales	Scotland	Northern Ireland
Living and Working Conditions	L.139/75	1365/75/EEC		Establishment of a foundation for living and Working conditions and financial provisions.			
	L.164/76	1417/76/EEC					

	EUROPEAN COMMUNITIES' LEGISLATION				UNITED KINGDOM LEGISLATION		
	Official Journal Reference	Principal Regulation/ Directive	Modification	Subject Matter	England & Wales	Scotland	Northern Ireland
Trading & Distribution of Toxic Products	L.307/74	74/557 EEC		Right of Establishment to trade and distribute toxic products transitional measures self employed users.			
Marketing of Fertilisers	L. 24/76 L.250/80	76/116/EEC 80/876/EEC		Marketing of High Nitrogen Content Ammonia Nitrate Based Fertiliser			
Safety Standards	59/221 66/45 L.187/76 L. 1/76	59/11 Euratom 66/3693 Euratom 76/579 Euratom		Basic safety standards for the health protection of the general public and workers against the dangers of ionizing radiation.	Medicines (Administration of Radioactive Substances) Regulations 1978. S.I. 1978, No. 1006.	Medicines (Administration of Radioactive Substances) Regulations 1978. S.I. 1978, No. 1006.	Medicines (Administration of Radioactive Substances) Regulations 1978. S.I. 1978 No. 1006.
Biological screening for lead	L.105/77	77/312/EEC		Biological Screening of the population for lead.	Control of Lead at Work Regulations 1980. S.I. 1980 No. 1248	Control of Lead at Work Regulations 1980. S.I. 1980 No. 1248.	

	EUROPEAN COMMUNITIES' LEGISLATION				UNITED KINGDOM LEGISLATION		
	Official Journal Reference	Principal Regulation, Directive	Modification	Subject Matter	England & Wales	Scotland	Northern Ireland
Safety signs	L.229/77	77/756/EEC		Safety signs at places of work	Safety Signs Regulations 1980 S.I. 1980 No. 1471.	Safety Signs Regulations 1980 S.I. 1980 No. 1471.	
	L.183/79		79/640/EEC				
Health of workers exposed to vinyl chloride monomer	L.197/78	78/610/EEC		Protection of the health of workers exposed to vinyl chloride monomer			
Health of workers exposed to chemical, physical and biological agents at work	L.327/80	80/1107/EEC		The protection of workers from risks related to chemical, physical and biological agents at work.			

	EUROPEAN COMMUNITIES' LEGISLATION				*UNITED KINGDOM LEGISLATION*		
	Official Journal Reference	Principal Regulation/ Directive	Modification	Subject Matter	England & Wales	Scotland	Northern Ireland
Dangerous Substances and Preparations	L.196/67	67/548/EEC		Classification and Labelling of Dangerous Substances	Packaging & Labelling of Dangerous Substances Regulations 1978 S.I. 1978 No. 209	Packaging & Labelling of Dangerous Substances Regulations 1978 S.I. 1978 No. 209	
	L.59/70		70/189/EEC				
	L.74/71		71/144/EEC				
	L.167/73		73/146/EEC				
	L.183/75		75/409/EEC				
	L.360/76		76/907/EEC				
	L.88/79		79/370/EEC		Packaging & Labelling of Dangerous Substances (Amendment) Regulations 1981 S.I. 1981 No. 792.	Packaging & Labelling of Dangerous Substances (Amendment) Regulations 1981 S.I. 1981 No. 792.	
	L.259/79		79/831/EEC				
	L.206/78	78/631/EEC		Classification and Labelling of Dangerous Substances (Pesticides)			
	L.88/81		81/187/EEC				
	L.189/73	73/173/EEC		Classification and Labelling of Dangerous Preparations (Solvents)			
	L.229/80		80/781/EEC				
	L.375/80		80/1271/EEC				
	L.303/77	77/728/EEC		Classification, packaging & labelling of paints, varnishes, printing ink, adhesives and similar products.			

	EUROPEAN COMMUNITIES' LEGISLATION				UNITED KINGDOM LEGISLATION		
	Official Journal Reference	Principal Regulation/ Directive	Modification	Subject Matter	England & Wales	Scotland	Northern Ireland
Lifting and Lifting appliances	L.335/73	73/361/EEC		Lifting and Lifting appliances			
	L.122/76		76/434/EEC				
Building and Civil Engineering Equipment and Machines	L.33/79	79/113/EEC		Determination of the noise emission of construction plant and equipment.			

	EUROPEAN COMMUNITIES' LEGISLATION				UNITED KINGDOM LEGISLATION		
	Official Journal Reference	Principal Regulation/ Directive	Modification	Subject Matter	England and Wales	Scotland	Northern Ireland
Tractors and Agricultural Machinery	L.84/74	74/150					
	L.262/76	76/763/EEC		Passenger Seats	Agricultural or Forestry Tractors and Tractor Components (Type Approval) Regulations 1975 S.I. 1975 No. 1475.	Agricultural or Forestry Tractors Components (Type Approval) Regulations 1975 S.I. 1975 No. 1475	Agricultural or Forestry Tractors Components (Type Approval) Regulations 1975 S.I. 1975 No. 1475.
	L.220/77	77/536/EEC	78/764/EEC	Rollover protection structures	Agricultural or Forestry Tractors and Tractor Components (Type Approval) Regulations 1979 S.I. 1979 No. 221	Agricultural or Forestry Tractors and Tractor Components (Type Approval) Regulations 1979 S.I. 1979, No. 221.	Agricultural or Forestry Tractors and Tractor Components (Type Approval) Regulations 1979 S.I. 1979, No. 221.
	L.179/79		79/622/EEC	Driver perceived noise level			
	L.105/77	77/311/EEC					
Electrical Equipment	L.24/76	76/117/EEC		Electrical equipment for use in potentially explosive atmospheres.	Agricultural or Forestry Tractors and Tractor Components (Type Approval) (Amendment) Regulations 1981 S.I. 1981 No. 689.	Agricultural or Forestry Tractors and Tractor Components (Type Approval) (Amendment) Regulations 1981 S.I. 1981 No. 669.	Agricultural or Forestry Tractors and Tractor Components (Type Approval) (Amendment) Regulations 1981 S.I. 1981 No. 669.
	L.43/79	79/196/EEC					
Pressure Vessels	L.262/76	76/767/EEC		Pressure vessels (framework directive)			

APPENDIX 4

HSE Improvement and Prohibition Notices

Improvement Notice: Form LP1

HEALTH AND SAFETY EXECUTIVE **Serial No.I**

Health and Safety at Work etc Act 1974, Sections 21, 23 and 24

IMPROVEMENT NOTICE

Name and address (See Section 46)	To ....
(a) Delete as necessary	(a) Trading as ....
	(b) ....
(b) Inspector's full name	one of (c) ....
(c) Inspector's official designation	of (d) .... Tel No. .... hereby give you notice that I am of the opinion that at
(d) Official address	(e) ....
(e) Location of premises of place and activity	you, as (a) an employer/a self employed person/ a person wholly or partly in control of the premises
(f) Other specified capacity	(f) .... (a) are contravening/have contravened in circumstances that make it likely that the contravention will continue or be repeated ....
(g) Provisions contravened	(g) .... The reasons for my said opinion are:- .... and I hereby require you to remedy the said contraventions or, as the case may be, the matters occasioning them by
(h) Date	(h) .... (a) in the manner stated in the attached schedule which forms part of the notice. Signature............................. Date.......................... Being an inspector appointed by an instrument in writing made pursuant to Section 19 of the said Act and entitled to issue this notice. (a) An improvement notice is also being served on .... of .... related to the matters contained in this notice.

Prohibition Notice: Form LP2

HEALTH AND SAFETY EXECUTIVE

Health and Safety at Work etc Act 1974, Sections 22–24 **Serial No.P**

PROHIBITION NOTICE

Name and address (See Section 46)	To ....
	....
(a) Delete as necessary	(a) Trading as ....
(b) Inspector's full name	(b) ....
(c) Inspector's official designation	one of (c) ....
(d) Official address	of (d) ....
	.... Tel No. ....
	hereby give you notice that I am of the opinion that the following activities,
	namely:- ....
	....
	....
	where are (a) being carried on by you/ about to be carried on by you/under your control
(e) Location of activity	at (e) ....
	involve, or will involve (a) a risk/an imminent risk, of serious personal injury. I am further of the opinion that the said matters involve contraventions of the following statutory provision:- ....
	....
	....
	....
	because ....
	....
	....
	and I hereby direct that the said activities shall not be carried on by you or under your control (a) immediately/after
(f) Date	(f) ....
	unless the said contraventions and matters included in the schedule, which forms part of this notice, have been remedied.
	Signature .... Date ....
	being an inspector appointed by an instrument in writing made pursuant to Section 19 of the said Act and entitled to issue this notice.

APPENDIX 5

Schedule 3 of the Health and Safety at Work etc. Act 1974

SUBJECT-MATTER OF HEALTH AND SAFETY REGULATIONS

1.—(1) Regulating or prohibiting—

(*a*) the manufacture, supply or use of any plant;

(*b*) the manufacture, supply, keeping or use of any substance;

(*c*) the carrying on of any process or the carrying out of any operation.

(2) Imposing requirements with respect to the design, construction, guarding, siting, installation, commissioning, examination, repair, maintenance, alteration, adjustment, dismantling, testing or inspection of any plant.

(3) Imposing requirements with respect to the marking of any plant or of any articles used or designed for use as components in any plant, and in that connection regulating or restricting the use of specified markings.

(4) Imposing requirements with respect to the testing, labelling or examination of any substance.

(5) Imposing requirements with respect to the carrying out of research in connection with any activity mentioned in sub-paragraphs (1) to (4) above.

2.—(1) Prohibiting the importation into the United Kingdom or the landing or unloading there of articles or substances of any specified description, whether absolutely or unless conditions imposed by or under the regulations are complied with.

(2) Specifying, in a case where an act or omission in relation to such an importation, landing or unloading as is mentioned in the preceding sub-paragraph constitutes an offence under a provision of this Act and of the Customs and Excise Act 1952, the Act under which the offence is to be punished.

3.—(1) Prohibiting or regulating the transport of articles or substances of any specified description.

(2) Imposing requirements with respect to the manner and means of transporting articles or substances of any specified description, including requirements with respect to the construction, testing and marking of containers and means of transport and the packaging and labelling of articles or substances in connection with their transport.

4.—(1) Prohibiting the carrying on of any specified activity or the doing of any specified thing except under the authority and in accordance with the terms and conditions of a licence, or except with the consent or approval of a specified authority.

(2) Providing for the grant, renewal, variation, transfer and revocation of licences (including the variation and revocation of conditions attached to licences).

5. Requiring any person, premises or thing to be registered in any specified circumstances or as a condition of the carrying on of any specified activity or the doing of any specified thing.

6.—(1) Requiring, in specified circumstances, the appointment (whether in a specified capacity or not) of persons (or persons with specified qualifications or experience, or both) to perform specified functions, and imposing duties or conferring powers on persons appointed (whether in pursuance of the regulations or not) to perform specified functions.

(2) Restricting the performance of specified functions to persons possessing specified qualifications or experience.

7. Regulating or prohibiting the employment in specified circumstances of all persons or any class of persons.

8.—(1) Requiring the making of arrangements for securing the health of persons at work or other persons, including arrangements for medical examinations and health surveys.

(2) Requiring the making of arrangements for monitoring the atmospheric or other conditions in which persons work.

9. Imposing requirements with respect to any matter affecting the conditions in which persons work, including in particular such matters as the structural condition and stability of premises, the means of access to and egress from premises, cleanliness, temperature, lighting, ventilation, overcrowding, noise, vibrations, ionising and other radiations, dust and fumes.

10. Securing the provision of specified welfare facilities for persons at work, including in particular such things as an adequate water supply, sanitary conveniences, washing and bathing facilities, ambulance and first-aid arrangements, cloakroom accommodation, sitting facilities and refreshment facilities.

11. Imposing requirements with respect to the provision and use in specified circumstances of protective clothing or equipment, including clothing affording protection against the weather.

12. Requiring in specified circumstances the taking of specified precautions in connection with the risk of fire.

13.—(1) Prohibiting or imposing requirements in connection with the emission into the atmosphere of any specified gas, smoke or dust or any other specified substance whatsoever.

(2) Prohibiting or imposing requirements in connection with the emission of noise, vibrations or any ionising or other radiations.

(3) Imposing requirements with respect to the monitoring of

any such emission as is mentioned in the preceding sub-paragraphs.

14. Imposing requirements with respect to the instruction, training and supervision of persons at work.

15.—(1) Requiring, in specified circumstances, specified matters to be notified in a specified manner to specified persons.

(2) Empowering inspectors in specified circumstances to require persons to submit written particulars of measures proposed to be taken to achieve compliance with any of the relevant statutory provisions.

16. Imposing requirements with respect to the keeping and preservation of records and other documents, including plans and maps.

17. Imposing requirements with respect to the management of animals.

18. The following purposes as regards premises of any specified description where persons work, namely—

(*a*) requiring precautions to be taken against dangers to which the premises or persons therein are or may be exposed by reason of conditions (including natural conditions) existing in the vicinity;

(*b*) securing that persons in the premises leave them in specified circumstances.

19. Conferring, in specified circumstances involving a risk of fire or explosion, power to search a person or any article which a person has with him for the purpose of ascertaining whether he has in his possession any article of a specified kind likely in those circumstances to cause a fire or explosion, and power to seize and dispose of any article of that kind found on such a search.

20. Restricting, prohibiting or requiring the doing of any specified thing where any accident or other occurrence of a specified kind has occurred.

21. As regards cases of any specified class, being a class such that the variety in the circumstances of particular cases within it calls for the making of special provision for particular cases, any of the following purposes, namely—

(*a*) conferring on employers or other persons power to make rules or give directions with respect to matters affecting health or safety;

(*b*) requiring employers or other persons to make rules with respect to any such matters;

(*c*) empowering specified persons to require employers or other persons either to make rules with respect to any such matters or to modify any such rules previously made by virtue of this paragraph; and (*d*) making admissible in evidence without further proof, in such circumstances and subject to such conditions as may be specified, documents which purport to be copies of rules or rules of any specified class made under this paragraph.

22. Conferring on any local or public authority power to make byelaws with respect to any specified matter, specifying the authority or person by whom any byelaws made in the exercise of that power need to be confirmed, and generally providing for the procedure to be followed in connection with the making of any such byelaws.

Interpretation

23.—(1) In this Schedule "specified" means specified in health and safety regulations.

(2) It is hereby declared that the mention in this Schedule of a purpose that falls within any more general purpose mentioned therein is without prejudice to the generality of the more general purpose.

INDEX